The Industrial Ethernet Networking Guide

Understanding the Infrastructure Connecting Business Enterprises, Factory Automation, and Control Systems

The Industrial Ethernet Networking Guide

Understanding the Infrastructure Connecting Business Enterprises, Factory Automation, and Control Systems

Donald J. Sterling, Jr.
Steven P. Wissler

THOMSON

DELMAR LEARNING

Australia   Canada   Mexico   Singapore   Spain   United Kingdom   United States

THOMSON
DELMAR LEARNING

The Industrial Ethernet Networking Guide
Donald J. Sterling & Steven P. Wissler

Executive Director:
Alar Elken

Executive Editor:
Sandy Clark

Senior Acquisitions Editor:
Gregory L. Clayton

Senior Development Editor:
Michelle Ruelos Cannistraci

Executive Marketing Manager:
Maura Theriault

Channel Manager:
Fair Huntoon

Marketing Coordinator:
Sarena Douglass

Executive Production Manager:
Mary Ellen Black

Production Manager:
Larry Main

Senior Project Editor:
Christopher Chien

Art/Design Coordinator:
David Arsenault

Editorial Assistant:
Jennifer Luck

Library of Congress Cataloging-in-Publication Data
Sterling, Donald J., 1951–
  The industrial Ethernet networking guide: understanding the infrastructure connecting business enterprises, factory automation, and control systems/ Donald J. Sterling, Jr., Steven P. Wissler.
    p.   cm.
  ISBN 0-7668-4210-X
  1. Manufacturing processes—Automation. 2. Production control—Automation. 3. Ethernet (Local area network system) I. Wissler, Steven P. II. Title.
  TS183.S753 2003
  370'.285—dc21   2002026792

NOTICE TO THE READER

Donald J. Sterling, Jr.

To the memory of my parents

Steven P. Wissler

*To my father, Paul Wissler,
who keeps running
the race of faith*

About the Authors

DON STERLING has been with Godfrey Advertising for 15 years and is currently director of technical strategic marketing. He has produced technical marketing materials for several of the leading structured cabling manufacturers. He is the author of several widely read books, including the recently published *Technician's Guide to Fiber Optics*.

STEVE WISSLER has worked for over 20 years writing and illustrating a wide range of marketing communications materials, including instrumentation, control, and industrial networking subjects. Since 1988, he has worked for Godfrey Advertising in Lancaster, Pa., where he is the creative director responsible for developing sales and marketing communications.

Contents

Preface

THE POINT

Most computers and printers in office buildings (and many homes) are connected using a networking standard known as Ethernet. This book is primarily about the industrial application of that standard, which is termed **industrial Ethernet (iE).**

The goal is to give the reader a comprehensive understanding of iE as a networking technology for factory automation and process control and also as an infrastructure enabling a wide range of business solutions—including enterprise resource planning (ERP), manufacturing execution system (MES), supply chain management (SCM), and customer relationship management (CRM).

Other books have been written on enterprise applications, e-business, Ethernet, network infrastructures, enterprise application integration, and control systems. We are indebted to these works and to the many standards organizations and vendors, who, in the spirit of the open Ethernet standard, make information publicly available over the Internet. But these resources tend to be a mile deep and an inch wide. Our approach is to offer a guide to the whole terrain of industrial Ethernet, from the factory floor to the top floor, to describe what is going on—and why.

We believe there is value in creating a broad, common ground for individuals and teams concerned with opening up factory networks to Ethernet and with connecting business enterprises to the factory.

On the manufacturing side, the rise of industrial Ethernet directly impacts control engineers and control system integrators, and, indirectly, plant managers responsible for operations management and production organization. On the business side, it impacts IT professionals, VPs of manufacturing and engineering, and CIOs, particularly executives interested in the competitive advantage gained by a common network infrastructure. And, of course, it interests equipment vendors, application developers, and solution providers promising to improve decision making, efficiency, agility, and profitability.

To create a common ground for this broad audience, technical jargon has been spelled out and concepts illustrated wherever possible. For those comfortable with jargon, however, this book provides a context for understanding how iE networking relates to the following:

Control-system, manufacturing-operations, and plant-management issues
- Migration from proprietary fieldbuses
- Network costing, design, modification, maintenance, and security
- Cellular and flow manufacturing

- Computer integrated manufacturing (CIM)
- Manufacturing execution system (MES)
- Advanced production scheduling (APS)
- Vendor managed inventory (VMI)
- Document management (PDM, CAD/CAM/CAE)
- Remote monitoring (wireless IP)

IT issues
- Enterprise-to-factory firewalls and network security
- Enterprise application integration (EAI)
- Implementation of ERP (enterprise resource planning), CRM (customer relationship management), and SCM (supply chain management) solutions

Executive issues
- Enterprise optimization involving finance, human resources, sales and marketing, and manufacturing
- Decision support systems
- Strategic collaboration between customers and suppliers

Each chapter contains a summary of key iE-networking concepts and a set of questions to assist the learning process. Readers who want specific answers to common problems dealing with physical connection, IP address configuration, and network design will find them in the appendices. Learning aids for this book are at the publisher's Web site address at www.electronictech.com. Vendor-specific information, research links, and reader discussions are available elsewhere (see www.ienguide.com). We invite readers to kindly draw our attention to any omissions or errors.

THE PARTS

Part One (Chapters 1 through 3) describes the data-carrying function of industrial Ethernet, the role of networking in industrial applications, and the business value of adopting industrial Ethernet.

Part Two (Chapters 4 through 8) examines the Ethernet standard. This part begins with a brief history of the technology and its operating features. Full-duplex operation, industrially hardened switches, connectors, topology, and QoS (quality of service) issues are discussed to show how Ethernet now meets the demands of the industrial environment.

Part Three (Chapters 9 through 11) shows how different protocols combine with Ethernet to provide various solutions—from a basic iE implementation to a complex protocol stack for distributed control systems. A table is provided to assist in making appropriate choices for existing and greenfield applications. Basic security methodology and technology are also discussed. Appendices A through C cover planning guidelines, IP addressing, subnet masking, and network examples. Appendices D through G contain a list of acronyms, a glossary, further resources, and answers to odd-numbered questions, respectively.

THE PEOPLE

The writing of this book spanned the "dotcom" craze, the "dotcom" bust, and the effects of a wartime economy. Yet we kept the faith that the community championing this dynamic technology—industrial Ethernet—would thrive.

Acknowledgments

The authors and Delmar Learning would like to thank the following reviewers:

Mark Fondl, President, Industrial Communication Technologies
Benson Hougland, Director of Marketing, Opto 22

Additional thanks go to the following people:

Raul Benoit, Lumberg
Richard B. Coan Jr., Controlled Environment Structures
Rick Downs, Opterna
Harry Furness, technical consultant
Robert Lowe and David Hendel, Loman Control Systems
Jack O'Brien, Capital Equipment Corp.
Norm O'Leary, Control System Integrators Association
Victor Wegelin, PMA Concepts

We especially thank our Godfrey colleague and director of integrated technology, Vince DiStefano, also owner of Cook Forest Online and coauthor of *Educator's Internet Companion* and *Child Safety on the Internet.*

TRADEMARKS

Efforts have been made to recognize trademarks for products mentioned in this book. All trademarks belong to the respective trademark holders. We apologize for any missed trademarks.

SUPPLEMENTS

An Instructor's Guide CD-ROM (ISBN 0-7668-4211-8) is available and includes all figures and answers to all questions from the textbook. Also included is a PC-based network sniffer program, a valuable tool in capturing and analyzing Ethernet frames and TCP/IP packets for an educational view into actual packet contents.

Donald J. Sterling, Jr.
Steven P. Wissler

PART ONE

Connecting Data, Factory Networks, and the Business Enterprise

A BEGINNING SCENARIO

It makes little sense to learn about a specific network technology without considering how networking in general improves information access so businesses save time, boost quality, cut costs, and increase profits. As an introduction, we will look at a scenario of a small company before and after it gets a network to see if it is better off.

The fictional business is Sally's Custom Soap Company. Without a network, handling information for even a single order creates a lot of work for Sally. She would rather spend her time doing research and development (R&D), but instead Sally and her employees spend about *three hours* per order coordinating personnel, juggling schedules, tracking down information, ordering supplies, and writing reports.

Even so, Sally's Custom Soap is surprisingly successful. Orders keep pouring in. Sally invested in computers and a production planning system to keep up, but there is still a lot of paperwork and confusion. Let us look at what she goes through to process a single order.

Life Before Networking

1. *Open job order.* Sally gets a phone call from a customer ordering 102 bars of Strawberry Sense soap. She enters the order into her production system as job number S1777 with a shipping date for three weeks later.
2. *Check inventory.* Sally walks to the warehouse to see if any bars of Strawberry Sense are in stock. Seeing only four, she assembles a job packet to start production on the remaining 98 bars. She looks up the recipe and types up the bill of materials.
3. *Check supplies.* Sally sees the recipe calls for olive oil, lard, tallow, palm oil, coconut oil, lye, water, vitamin E oil, red coloring, dried strawberries, and strawberry fragrance extract. She walks over to the storeroom and notices she needs to order more coconut oil and vitamin E oil to complete the batch. She calls her supplier and schedules delivery for the next morning.
4. *Production entry.* Sally walks the job packet over to the production manager—her sister, Sarah—in the blending room. The manager puts the job in a queue so production is ready to begin tomorrow when the supplies arrive.
5. *Tracking supplies.* The next morning, Sally calls Sarah to see if order S1777 is in process as scheduled. No, it's not, because the oils still haven't arrived. Sally calls the supplier, who apologizes and promises delivery early in the afternoon.
6. *Receiving department.* When the supplier delivers the oils, the receiver walks the invoices over to Sally. Sally then calls Sarah to begin production.
7. *Feed stock.* The manager loads the oils and other materials onto a cart and pushes it over to the worker assigned to prepare the batch.
8. *Processing.* The worker follows the recipe. She carefully combines lye with hot water and lets the mixture cool to about 98°F. The oils are warmed to about the same tem-

perature. When the lye water and oils both reach 98°F, the lye water is poured in a thin stream into the oils, which are constantly stirred. When the mixture reaches trace—about the thickness of cake frosting—the fragrance and dried strawberry bits are added. The mixture is then poured into moulds to make 98 bars and cured for 24 hours.

9. *Quality control.* The next day, Sarah calls Sally to check the soap quality. Sally walks to the production area, retrieves the recipe and job order from the packets. She looks at the bars and likes what she sees and smells. The bars are moved to a temperature- and humidity-controlled room for three weeks of curing.

10. *Packaging and shipping.* Three weeks later, Sally's production system informs her that order S1777 is ready for wrapping, boxing, and shipping. Sally prints out the shipping information and walks it over to the packaging department. The packager pulls the 98 bars out of the curing room and four bars out of inventory to complete the order. The packager completes the UPS shipping label and invoice and walks copies over to Sally.

11. *Invoicing and billing.* Sally photocopies the shipping invoice, places one copy inside the completed job jacket, sends another to accounts payable, and encloses yet another in the customer's bill. She enters the ship date into her production program, which marks the job as complete.

EVALUATION
One job involves six walking trips, eight documents, and eight phone calls, plus a delay over supply delivery (Figure I–1).

Frustrated about the wasted time, Sally talks to a supplier, who says his information technology (IT) and process control experts came up with a solution for him. Sally learns that it will take a team composed of an outside control systems integrator and a business application provider, plus her own accountant, plant manager, engineers, and warehouse manager to build a solution. Together, the team maps out the kinds of applications, databases, workstations, instruments, control devices, and control network that will meet the business's growing needs. Within sixty days, a new system has been installed and debugged. Now let us look at the same scenario with an industrial Ethernet network in place.

Life with a Network

1. *Open job order.* Sally gets an e-mail from a customer ordering 102 bars of Strawberry Sense soap. She enters the order into her resource-planning system as job number S1777 with a shipping date for three weeks later.

2. *Check inventory and supplies.* Because shipping now uses bar coding to keep track of inventory, the resource-planning system pulls data from a warehouse data management program to show that 4 bars of Stawberry Sense are in stock. The planning system calculates only 98 bars need to be produced. Sally calls up the soap recipe on the planning system and compares it with the material on hand. She sends an e-mail

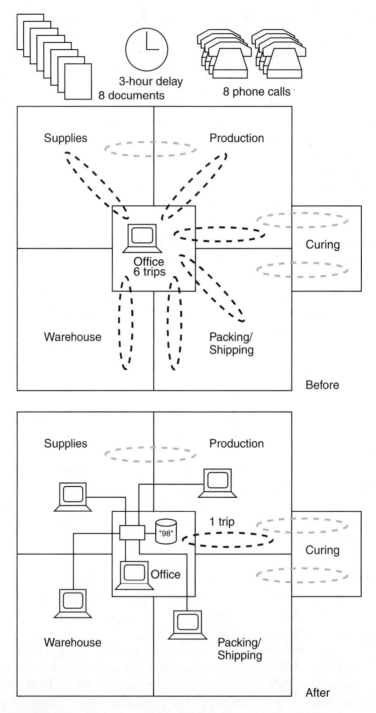

FIGURE I–1   Scenario

to the supplier to ship two missing ingredients—coconut oil and vitamin E oil—tomorrow morning. The vendor's supply management system automatically sends a reply e-mail that the materials will be delivered in the afternoon, not in the morning.

3. *Production entry.* Sally enters the job order and supply status into a manufacturing-execution program to schedule production tomorrow.

4. *Receiving department.* The next afternoon, the supplier delivers the oils. The receiver enters them into the inventory system, which automatically updates the bill of materials list in the planning resource systems.

5. *Feed stock.* The next afternoon, the worker on the production floor looks at the factory scheduler and sees that 98 bars of Strawberry Sense soap need to be made. She clicks on the order, looks at the recipe, and notes that all the ingredients are in the supply room. The worker loads the oils and other materials onto a cart and pushes it over to the processing vats.

6. *Processing.* The worker follows the recipe. She carefully combines hot lye into a vat with hot water. A Web page generated by her data acquisition system displays the temperature as 98°F. In another vat, the oils are being heated. When both vats' temperatures reach 98°F, the lye water is poured into the oils, fragrance is added, and the mixture is poured into moulds. The worker enters the number of bars produced—98—and the time the job is finished. She also saves the process temperature data in a database. All information about who, what, when, where, how many, and how long is now accessible through the planning resource system.

7. *Quality control.* The next day, the planning system alerts Sally to check the bars for quality. Sally goes to the worker's station and pulls information from the system on the recipe, the quantity, delivery, and production time needed to make the 98 bars. Sally likes what she sees and smells, and congratulates the worker on a job well done. Then the bars are taken to the curing room for three weeks.

8. *Packaging and shipping.* After three weeks, the planning system alerts shipping that order S1777 is ready for wrapping, boxing, and shipping. The packager pulls the 98 bars out of the curing room and four bars out of inventory to complete the order. The system automatically completes the UPS shipping label and invoice.

9. *Invoicing and billing.* Sally's system shows the job has been shipped. She attaches the shipping invoice and encloses it in a bill e-mailed to the customer. The system notes that the bill has been sent and marks the job as complete.

EVALUATION

Because the Internet connects Sally and her production systems to her customers and suppliers, she saves phone calls, paperwork, and confusion. She also saves trips to the plant floor, because the factory network provides schedules, recipes, and control data to workers and systems as needed. As a result, Sally only takes one trip for quality control purposes. Thanks to efficient information flow, Sally has been able to reassign her sister Sarah to help with R&D.

Basic Business Advantages of Networking

In the last scenario, networking benefits the operation by reducing the informational friction between people and processes. There are many savings obtained through networking—savings that occur whenever the lag between information and action can be shortened or removed.

This scenario provides a context for the detailed discussion that follows. You will note that the number 98 also shows up in the following chapters to illustrate how industrial Ethernet (iE) networks handle the two types of data common in industry, namely: (1) numbers for a specific quantity of units, and (2) numbers representing a continuous value, such as temperature readings.

While applying networking technology in the real world is complex, the basic idea remains simple: to replace the friction in work due to inefficient exchange of information with the speed, accuracy, profitability, and creativity that results from good communication.

The Ethernet Communication Evolution: A New Framework for Industrial Networking

1

OBJECTIVES

After reading this chapter, you will be able to:

- Understand the basics of Ethernet and factory networking as an introduction to the detailed discussion in following chapters.
- Dissect an Ethernet data packet and understand how the parts relate to Internet protocols.
- Know how a piece of data travels from its point of origin to a destination application.
- Learn the importance of encapsulation and layering in data transmission.
- Understand how developments in information, business, factory control, and networking technologies pushed Ethernet to the forefront of industrial networking.

Industrial Ethernet (iE), as discussed in this book, is simply the implementation of Ethernet technology to carry information between points inside and outside a manufacturing plant or process facility.

There are no deep, dark secrets that separate Ethernet from industrial Ethernet. That is because iE is a subset of the same Ethernet standards used to create the communication connections—the **network**—between computers, workstations, and related devices.

Networks are classified either as **local area networks (LANs),** which are located in a limited geographical area such as an office or factory, or as **wide area networks (WANs),** which are located outside the facility but connect with the LAN to provide necessary communications. Together, LANs and WANs comprise the combined business network known as the **enterprise.**

In offices, an Ethernet LAN is taken for granted as part of the basic business **infrastructure.** In other words, it is a basic means of conveyance that everyone uses and abuses, like the highway system.

In many industrial plants, however, the LAN is a relatively small path that connects computers and workstations used by managers and engineers. Other roads, namely non-Ethernet **industrial networks,** are usually kept isolated from the enterprise. Industrial networks are dedicated to conveying critical control information and operational data to operators, equipment, controllers, valves, and sensors.

Before Ethernet's popularity, industry developed these robust networks to communicate control data, which is far more critical than office documents or e-mail. While the goal of Ethernet technology is the same in shop-floor and top-floor networks—to ensure timely, reliable data delivery—the stakes are much higher in industrial applications.

In the industrial environment, network outages or overloads can jeopardize production, property, and people's lives. Factory networks must reliably and quickly transmit temperature, pressure, and position data from sensors, actuators, and controllers to avoid problems in the process or operation. They must also transmit programs, recipes, diagnostics, and operational data between **programmable logic controllers (PLCs)** and PC workstations to ensure equipment is operating within proper parameters (Figure 1–1).

In recent years, the demand for instant and universal access to data from the plant floor has increased dramatically. Consequently, the role of the industrial plant LAN is being expanded. In addition to connecting PLCs and manufacturing execution systems used by plant managers and supervisors, it is also connecting with business information and planning systems used in the business headquarters, as well as with information systems used by suppliers and customers. At the same time, the Ethernet infrastructure is being extended downward to connect with critical sensors and devices on the plant floor.

It is the continuing evolution of Ethernet for handling a wide range of business activities, including plant automation and control, which makes learning about industrial Ethernet worthwhile.

THE LEARNING PROCESS

Because industrial Ethernet touches on many technical and business areas, there are a number of ways to dive into the subject. This book uses technical standards, communication models, data dissection, component interaction, along with case histories in the appendices, to satisfy the interests of various readers.

Technical Standards Approach: IEEE 802.3 View

Network-engineering, electronic, and information professionals—and any reader who is a glutton for punishment—can study the Ethernet standard published by the Institute of Electrical and Electronic Engineers (IEEE).

Ethernet is an **open standard,** which means Ethernet procedures are developed and disseminated to the public either for free or for a nominal fee. The IEEE deserves a word of "thanks" for offering their Ethernet standard, *802.3 CSMA/CD Access Method,* as a *free* Internet download. (See www.ieee.org for details.)

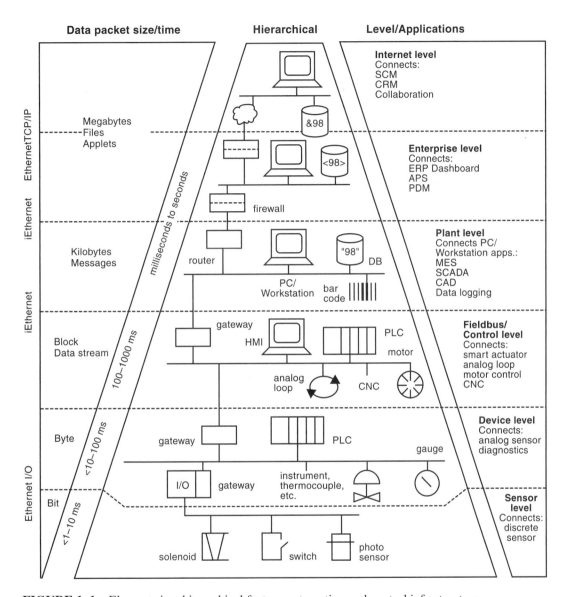

FIGURE 1–1 Elements in a hierarchical factory automation and control infrastructure

Another professional-level resource is the Internet Engineering Task Force (IETF). The IETF is an international community of network designers, operators, vendors, and researchers concerned with developing the network of public networks, known as the **Internet.**

IETF working groups and others contribute **Requests for Comments (RFCs),** which are technical notes that discuss relevant aspects of computer networking, including procedures, programs, and concepts. RFC specifications make it possible to use Ethernet for Internet communication. RFCs can be accessed at www.ietf.org.

Communication Models Approach: OSI-Layer View

One of the challenges of writing about Ethernet is to avoid letting complex technologies—which are necessary to make Ethernet a useful tool for industrial networking—obscure its basic simplicity.

A good way to understand Ethernet is to look at it as a conceptual model. For example, it is easy to show how Ethernet functions in a network by using the image of a *multilayer stack.*

This multilayer model is not just a metaphor. In fact, a layered reference model was developed in 1978 by the International Organization for Standardization (ISO—see www.iso.ch). It is a seven-layer stack known as the Open Systems Interconnect Reference Model (OSI/RM) and is usually referred to as the **OSI model** or **seven-layer model.** The OSI model is the basis for the U.S. Government's own standard for network environments, Government Open Systems Interconnection Profile (GOSIP), which was published in August 1990.

It is important to remember that the OSI model is a template, nothing more. It is useful for comparing all kinds of network communication standards. Each layer in the OSI model details the specific functions that a standard *ideally* uses for network communications. Strictly speaking, Ethernet comprises just the first and part of the second OSI layers.

Data Dissection Approach: Packet View

Because of its conceptual importance, the OSI model is frequently referred to throughout this book and is the basis for dividing Part Two and Part Three. But as previously mentioned, the OSI model is only a template. It does not show what is really happening in Ethernet network communications. For this, it is necessary to look into the actual code of an Ethernet transmission.

Ethernet can be compared to the system of roadways, addresses, vans, and mail bags used to deliver letters house to house. Consequently, it involves physical components and numerical codes that enable devices to connect and to transmit digital data.

As in postal mail handling, specific processes and numbers are involved in wrapping, padding, sealing, addressing, sorting, transporting, tracking, delivering, and responding to messages.

Ethernet uses particular rules, known as **protocols,** to create a basic digital container or envelope, known as a **frame,** to enclose all the other codes and data necessary for network communication.

Component Interaction Approach: System View

This chapter begins with dissecting the contents of an actual Ethernet transmission. Then it considers how the contents are handled by hardware and software components during different stages of the network-communication process.

While a bit more complex than starting with the OSI model, beginning with the code in an Ethernet frame and showing how it is used by hardware and program applications is probably a better introduction.

That is because it is easier to grasp what is tangible and real. In fact with the tools mentioned in this book, you can get your hands on a data packet and see the contents for yourself!

This approach also demonstrates that in the digital world, every interaction must be programmed in code. Beginning with an actual message is a reminder that industrial Ethernet is not just a faster and cheaper way to communicate data. Rather the central value of iE networking, as in any other form of communication, is its ability to convey content, context, and insight.

DATA DISSECTION

In this section, we will be dissecting a **data packet,** which is a self-contained unit of code and data that encloses all the addressing information necessary to reach its destination. A data packet refers to any block of code delivered over the network. Data packet is sometimes used interchangeably with datagram. Technically, however, a **datagram** refers to a specific unit formed by one of the protocols discussed below.

Packet Types

Packets of different sizes, compositions, and functions are formed by particular protocols throughout a communication session. As will be shown in Part Two, each protocol plays a specific role at every stage of the request-and-reply communication exchange. A single packet may contain, for example, a small amount of code that requests the address of the network interface card (NIC) inside the recipient's PC. It contains no information of interest to the end user, unless you are a network troubleshooter.

To be relevant to more readers, the packet shown here is a reply message containing data of interest to the end user. Specifically, we will be considering a single Ethernet data packet transmitting the temperature value 98 over the Internet (Figure 1–2). Control data is not often communicated over the Internet, because, as the example shows, it takes a large amount of code simply to display the temperature value 98 on a Web browser. But because the Internet is being employed more frequently, it is valuable to consider it from the start.

Keep in mind that in reality a data packet is transmitted as electrical pulses representing a continuous string of binary code, as shown in the figure. Markers have been added to show how the code is divided into functional parts of prescribed lengths.

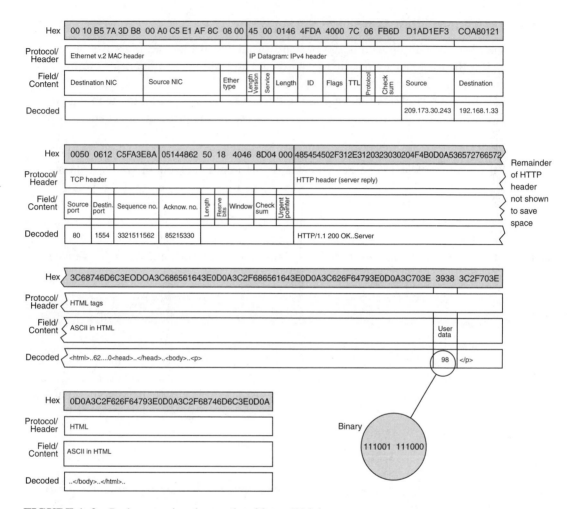

FIGURE 1–2 Packet carrying the number 98 to a Web browser

Packet Parts

HEADERS

A **header** is simply a segment of code that is added to the front of another segment of code known as the **payload** (Figure 1–3). The header contains the **protocol control information** that allows the payload to be properly processed.

All digital messages sent to controllers, digital phones, or personal computers begin with a header. Depending on the protocol used to form the header, the control information may be comprised of destination and origination addresses, protocol type, format, length, and other packet or network information. Even when the packet contains data of value to the user, the number of headers in the packet involves far more code than the user data itself—as seen in Figure 1–2.

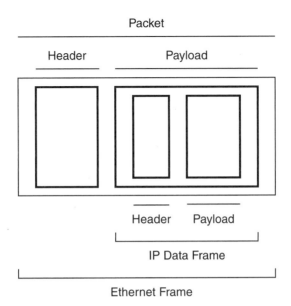

FIGURE 1–3   Relationships between headers, payloads, packet, and frames

FRAMES

The header and payload are contained together within a frame—the distinctive structure or format that serves as a container for the code. Specific protocol information or even other frames are contained in designated fields within a frame. The frames function together as a sequence of digital envelopes; one envelope or frame is nested, or **encapsulated,** inside another.

For example, Figure 1–4 shows how the code for the HTML page containing the user data is itself contained within an HTTP frame, which is contained within a TCP frame, which is contained within an IP frame, all of which are sandwiched inside the Ethernet frame. Each frame can be opened and processed (**decapsulated**) without disturbing the other frames nested within, because each header provides a separate layer of protocol control information to allow the payload to be processed independently.

Packets, headers, and frames can be easily confused with nomenclature for layers and stacks, so let us take a moment to clarify.

As previously mentioned, a **packet** is a complete unit of code that contains all the information necessary for transmission. Technically known as the **protocol data unit (PDU),** a packet is comprised of two parts: a **header,** which is technically called the **protocol control information (PCI),** and the **payload,** also known as the **service data unit (SDU),** which contains encapsulated code and data that are processed as a unit according to the protocol control information. Because the header and payload are contained in a rigid format, the entire structure is known as a frame. So what is the difference between a frame and a packet?

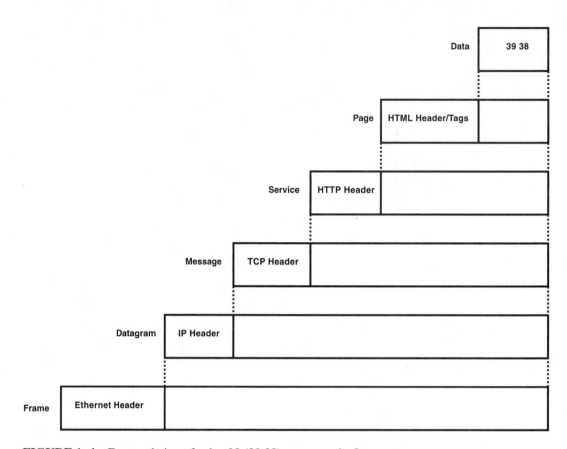

FIGURE 1–4   Encapsulation of value 98 (39 38) sent over the Internet

A **frame** refers to the structural container of a packet. In the case of an **Ethernet frame,** it refers to the largest possible structure that contains all the headers and payloads, which is the equivalent of a data packet, as shown in Figure 1–3. It may also refer to substructures, such as the IP data frame. As you will see, a **message** transmitted over a network is comprised of many packets arranged in a numbered sequence, while a packet is comprised of many frames arranged by encapsulation.

Layers and **stacks** refer to the type of protocol control information used to create the headers and frame. **Communication stack** describes a group of protocols in related layers. For example, the **TCP/IP stack** stands for the transmission control protocol and the Internet protocol, respectively, employed together in the transport and network layers to enable Internet communication.

As mentioned before, layer refers to the general network-communication function performed by a protocol or suite of protocols. Layering ensures that at each stage of the communication process, the proper protocols are being observed. The data packet in this chapter

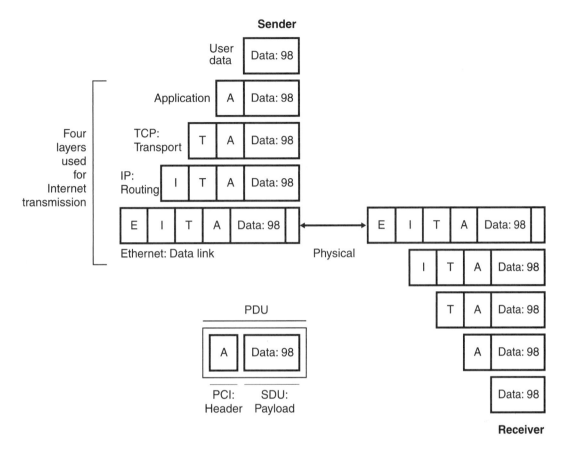

FIGURE 1–5   Headers related to functional layers

has worked its way up through four layers of protocols—data link, network, transport, and application layers—before making it to the user's Web browser at the user layer (Figure 1–5).

Packet Contents

The packet contents in Figure 1–2 show a single reply packet containing data of interest to the user. It contains the number 98, which will eventually be displayed by a Web browser. But before that happens, several different headers must be processed. Since the headers are in code, it is helpful to look at binary and hexadecimal encoding before dissecting the header.

BITS AND HEXADECIMAL CODE
The actual data read by the computer are binary digits, the most basic element of the data packet. **Binary digits** are represented by two numbers—0 and 1—each of which is known as a **bit.**

98 in ordinary form	9 in ASCII hexadecimal form	8 in ASCII hexadecimal form	9 in binary form	8 in binary form
98	39	38	111001	111000

FIGURE 1–6   Encoding the ordinary number 98 for HTML presentation

To make life easier for network analysts who must read packet contents, a **packet analyzer** or **sniffer** program is a helpful tool. A packet analyzer performs a **dump** that translates binaries into more readable forms.

Ordinary numbers, like the number 98 in our example, are encoded into binary and **hexadecimal** forms based on the **American Standard Code for Information Interchange (ASCII),** as shown in Figure 1–6.

Hexadecimal code is based on bits being arranged into a four-bit group known as a **byte.** Within each byte, the binary digits can be arranged sixteen different ways, which is why it is called **hexadecimal.** Each combination is assigned a unique value. Because there are over ten possible combinations, a number from 0 to 9 is used to represent the first ten combinations, then a letter of the alphabet from A to F is used for the six remaining combinations.

The ASCII standard determines what code values are used to represent ordinary numbers and letters. Note that the presentation of numbers on a Web page is based on the individual values of the number (9 and 8 separately), not the ordinary value (98). Thus, the ordinary number 9 is encoded as hex 39 and the number 8 is encoded as hex 38, as seen near the end of the data packet. But remember, this is a dump of the binary code. What was actually transmitted over the computer are electronic signals representing the binary code for 39 (111001) and 38 (111000).

HTML HEADER

The hex values 39 38 are bracketed on either side with **Hypertext Markup Language (HTML)** code, known as **tags.** The dump shows the tags as hex values (which can be converted back into readable HTML tags by using an ASCII conversion table. HTML tags are used by the Web browser to determine the size, position, and font of the numbers being displayed. HTML follows standards set forth by the World Wide Web Consortium (W3C).

HTTP HEADER

The **Hypertext Transfer Protocol (HTTP)** header performs application-layer functions by making it possible for HTML files to be exchanged over the Internet. HTTP standards are edited by the IETF and codified in RFC 2616.

TCP HEADER

Transmission Control Protocol (TCP) handles the delivery of the HTML file from the **server,** the source device replying to the request made by the **client,** the destination device. TCP performs transport-layer functions by establishing and maintaining the connection between the transmitting computers. It also numbers packets in numerical sequence to detect lost packets and to trigger retransmission. The TCP standard is published as RFC 793.

IP HEADER

Internet Protocol (IP) performs network-layer functions by adding the proper network address to each data packet as it moves between different network devices (NICs, routers, switches, etc). These header and payload units are known as **datagrams.** A network address is a four-byte number derived from a scheme developed by the **Internet Assigned Numbers Authority (IANA).** The IP standard is set forth in RFC 894.

ETHERNET HEADER

The Ethernet header performs data-link-layer functions by associating the network IP address with the physical address of the **network interface card (NIC).** A chip on the circuit board of each NIC used in the communication session contains a unique **media access control (MAC)** number assigned by IANA. The Ethernet header contains those two addresses and the specific type of Ethernet frame in front of the payload that contains all the other headers and data. Details on the Ethernet header and the data-link layer to which it conforms are discussed in Chapter 4.

Compared to other headers, the Ethernet header has a scant number of subsections or **fields** of code. The first five bytes identify the destination computer's NIC; the next five bytes identify the NIC card that is the source of the transmission; and the last two bytes identify the Ethernet protocol type. This lean header reflects Ethernet's limited, but critical double role of:

1. Creating a frame to contain the other protocol control information that establishes and maintains a communication session.
2. Providing a common standard of voltages and amperage and other physical components necessary to transmit the data as an electrical signal.

COMPONENT INTERACTION

The previous anatomy lesson showed the headers involved in a single reply message. The Ethernet frame was the shipping envelope. When cut open, it reveals the TCP/IP, HTTP, and HTML headers that wrapped the number 98 for delivery.

To get a clearer picture of an Ethernet communication transaction, it helps to see how the anatomy of a data packet interacts with applications and hardware.

Figure 1–7 shows the relationships among the packet, programs, and hardware in Microsoft Windows 95 and Windows NT computers. These operating systems include the software to provide specific layer functions, known as **services,** which form the packet and host the communication session.

Important fields	Hardware and software components employed	Layer
HTML: ASC II hex code: 39 38	Win-sock compliant application: Internet Explorer	User layer
HTTP		Application layer
TCP port number and IP address	Winsock.dll (forms a session by creating a socket by associating or binding the application's port to the packet's IP address)	Session layer
TCP header	TCP/IP stack	Transport layer
IP header		Network layer
MAC header	Network card	Data-link layer
Not applicable		Physical layer

FIGURE 1–7   Relationships among packet fields, Windows OS components

Software

A **stack** is created by software in the operating system that arranges layers of code involving port numbers, IP addresses, and MAC addresses in accord with appropriate protocols.

PORT NUMBER

Before the number 98 can be displayed by the browser, the data packet must find the application. Finding an application is a matter of finding the port servicing that application. A **port** is the end point of the connection where the data leaves the packet and enters the application. Ports are established electronically in the computer's logic; otherwise a physical wire would be needed for each application. Many logical ports can be opened simultaneously to serve as sending and listening channels. While ports make it possible to use just one network cable for all kinds of communications, they create numerous security problems, as discussed in Chapter 11.

Each application is assigned a port number. For example, a Web server listens for HTTP requests on port 80. Numbers under 1024 are "well-known port numbers" that are assigned by the **Internet Corporation for Assigned Names and Numbers (ICANN).** Up to 65535 ports are available in total.

SOCKETS

A socket associates or **binds** the IP address and the port number to create the connection end point. In this case, the reply data packet is being transmitted to the machine at IP address

192.168.1.33, and the Web browser will process information being channeled through port 1554. The socket creates a **pipe,** a one-way communication link that allows information to be passed from the server's application program to the client's browser on port 1554, while information from the browser is directed to the server on port 80, as seen in Figure 1–2.

A **socket** is created by a separate computer program that reads and writes the protocol control information into the proper header-payload format. Sockets are used whenever a client or server program application requests a network service.

Socket designs vary with the system. For example, the Microsoft **Winsock** standard was derived from the original Berkeley UNIX interface for sockets. In the case of a Windows NT computer, the browser makes a request to the Winsock **application program interface (API),** which runs the subroutine: winsock32.dll. **Winsock** is responsible for getting TCP/IP address and other protocol control information into the data packet.

Hardware

DEVICE COMPONENTS

NICs.   Network interface cards are the printed circuit boards installed in PCs, workstations, or other devices. They provide the physical and electrical connections between the network cable and the system. Each NIC has at least one connection port where the network cable is attached. Only one physical port may be used at a time.

Firmware.   Winsock acts in concert with the NIC by passing the TCP/IP stack to the NIC's **driver,** which is a small program that handles the input and output request between the system and the hardware. **Firmware** is a program stored in the NIC's memory chip that processes the TCP/IP stack and provides the distinctive six-byte MAC address that gets encoded in the Ethernet header.

MAC addresses.   In accord with Ethernet protocol, the NIC establishes the physical connection *and* supplies the device's MAC address to link the data frame to the receiving device's MAC address. Therefore a NIC operates at the physical and the data-link layers. While a detailed discussion of sockets, device drivers, and the TCP/IP stack are beyond the scope of this book, Ethernet physical and data-link layers—including cabling—are discussed in Part Two.

Encoding chip.   Before leaving the computer, the code is in a binary bit stream form. Before exiting the computer, this bit stream must be processed by a **Manchester encoding chip** that turns it into the proper electronic waveform to travel over the network cable in most 10 Mb/s-Ethernet applications. High-speed (100 Mb/s) and some versions of optical Ethernet use more complex encoding.

NETWORK COMPONENTS

Cabling and topology.   The data packet leaves the NIC via the port connected to the cable (or wireless transmitter) and follows the layout of the cabling, known as the network **topology.**

When connecting devices that can both originate and receive messages in **peer-to-peer** communication, either a **ring** or **bus** topology is employed (Figure 1–8).

In **ring** topology, the connected devices form a continuous chain in a closed loop. In **bus** topology—often used in factory automation and control networks to connect sensors and controllers with PLCs, supervisory control and data acquisitions (SCADAs), and supervisory PCs—all devices are connected to a single main cable. (Bus topology should not be confused with **fieldbus,** which refers to the bidirectional protocols used in device-level networks connecting instruments and controllers.)

In a **client/server network**—typically used in offices or other settings that rely on a centralized computer for file sharing—a **star** topology is used. The term *star* refers to communication emanating from a central point, in this case, the server. A dedicated gateway device can also serve as a central connection point, allowing the server to function as a peer among peers.

Peer-to-peer communication is the basis of iE LAN design. It gives a device the freedom to function either as a client or server, depending on the application. In our example, the number 98 was embedded in a Web page generated by a smart device and sent to a Web server, where it could be accessed by a remote PC over the Internet. The page could have just as easily been accessed directly from the smart device, with the server playing a standby role for purposes of data logging.

Linking devices.   In any kind of topology, **linking** hardware—such as **switched hubs, routers,** or **proxy servers**—are used to direct the packet to different networks or specially created **subnetworks (subnets).**

As the packet travels from device to device, it is said to hop, that is, to take a very brief stop at each addressed device, known as a **gateway node,** before being sent to the next point. What occurs during each stoppage? It depends on the device:

- *Hub.* Simply passes the packet through.
- *Switch.* Reads network address (MAC or IP) and passes the packet to proper destination.
- *Router.* Reads IP address and port number in TCP header and passes packet to proper subnet.
- *Gateway.* Reads application header and translates data into form suitable for a different type of physical network.

Because the packet is self-contained, it generally remains intact from hop to hop—except after passing through a gateway. Because a gateway connects two networks using different physical media, such as Ethernet and a serial modem telephone line or a **digital subscriber line (DSL)** modem line, the packet may have to be downsized to fit the other network's **maximum transmission unit (MTU)** limit. The gateway's IP protocol breaks the datagram into smaller parts—a process known as **fragmentation.** An identification number is inserted into the IP header of each fragment so the datagrams can be reassembled into the original packet. Then the rebuilt packet is passed up to TCP's services, where it hops back onto Ethernet media connected to the NIC.

Finally, the packet reaches the end point, the **host node,** which is the destination computer.

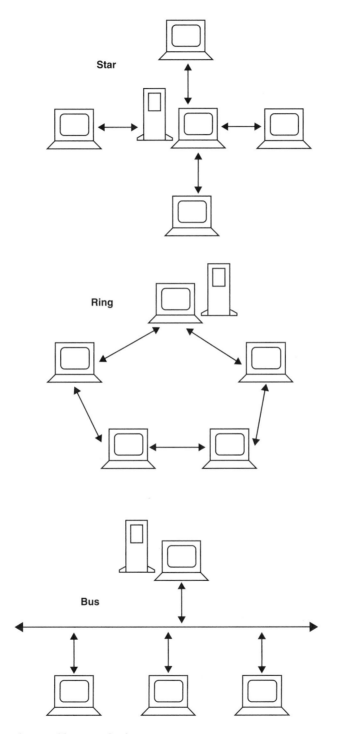

FIGURE 1–8   Star, ring, and bus topologies

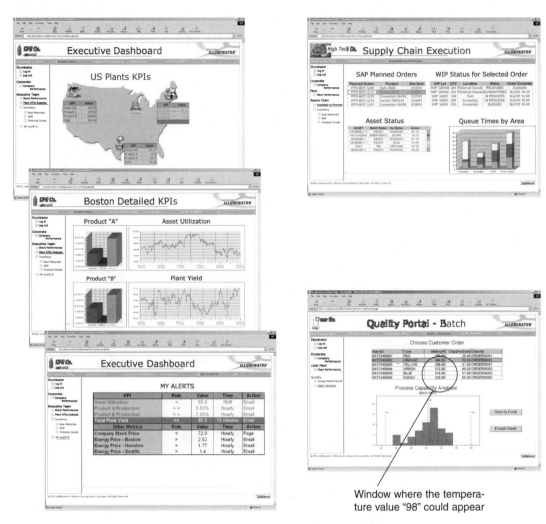

Window where the tempera-
ture value "98" could appear

FIGURE 1–9   Browser-based executive dashboard, supplier, and quality portals
(Courtesy of Lighthammer)

Application Components

WEB BROWSER

As the packet reaches the host NIC's firmware and moves up through the Winsock API, the payload is extracted during decapsulation and passed to the next layer, ultimately reaching the application layer. The HTTP header is processed by the Web browser, which reads and displays the HTML tag's ASCII code. At last, 98 shows up on your desktop computer as shown in Figure 1–9.

The universality of the Web browser makes it the logical information interface for users across the enterprise, whether on the plant or the executive floor. Consequently, many PLC

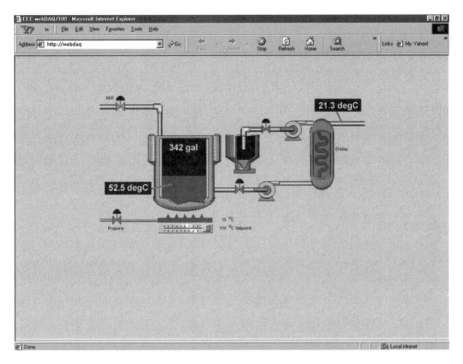

FIGURE 1–10   Process view created on a Web-enabled DAQ system
(Courtesy of CEC)

manufacturers now offer an Ethernet connection option, plus embed TCP/IP into devices to provide Internet connectivity. With Internet-ready PLCs, data can be extracted from plant-floor databases, collected on a Web server, and displayed for decision makers wherever they may be.

DATABASES
In addition to browser display, the number 98 can be reported, graphed, and stored by other applications, such as accounting software, **enterprise resource planning (ERP)** systems, or databases that may reside on older, so-called **legacy,** mainframe computer systems.

SCREEN SCRAPING
Unfortunately, a Web browser does nothing to get useful data into business applications. It is sometimes necessary to employ **screen scraping** programs to convert unstructured HTML data into row (record)/column (field) coordinates used in spreadsheets and databases. Practically speaking, the Web browser and HTML are better at presenting big-picture events in an operation, like graphically displaying tank levels, set points, and trouble spots, as shown in Figure 1–10.

DATA INTEGRATION

To minimize the need for expensive data conversion programs, an alternative to HTML was developed—**Xtensible Markup Language (XML).** Data with XML tags can be displayed in browsers and human-machine interfaces (HMI) panels and data historians used in plant control. But XML-tagged data also can be easily integrated with databases on back-end business systems, such as Oracle. Consequently, XML is an important technology in e-commerce, supply-chain management, and trading exchanges applications where different users need easy access to the same data.

XML is an open standard edited by the W3C. Various versions, such as factoryXML, are being developed to handle data specific to factory automation and control.

For example, here are the HTML tags used to display the value 98:

```
<html>
        <head></head>
        <body>
                <p>98</p>
        </body>
</html>
```

If XML were used, the tags could look like this:

```
<?xml version="1.0" standalone="yes"?>
        <Workstation_1>
                <TankTemperature>
                        <Pressure>98</Pressure>
                        <Units>Fahrenheit</Units>
                </TankTemperature>
        </Workstation_1>
```

Notice how the number 98 is embedded in tags that establish the meaning of the information, not just its appearance on the page.

APPLICATION PROGRAMS

In place of Web browsers, operators and managers on the plant floor handle data using dedicated monitoring and recording technologies:

1. *Expensive, proprietary HMI panels.* Designed specifically to display information pulled from the control network to let the operator make appropriate adjustments. The data is then stored on a proprietary data historian database.
2. *Less expensive, off-the-shelf control software.* Requires laborious manual **data mapping** to assign data transmitted over a specific wire to its proper place in the HMI and database application. Today, object-oriented programming environments employ

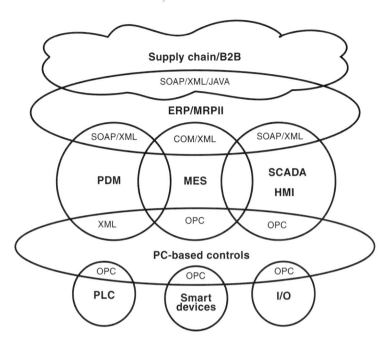

FIGURE 1–11   Software tools that link data between applications

reuseable modules for APIs and drivers to make it easier to write applications that exchange data between applications, as detailed in Chapter 10.

Here it is worth briefly introducing **object linking and embedding (OLE) for process control,** known simply as **OPC.** This Microsoft Windows-based technology is often used to tag, organize, and link device-level data to HMI, **supervisory control and data acquisition (SCADA),** and custom control applications (Figure 1–11). With OPC, field devices can serve data to the client, such as a controller or HMI, and can act as a server to an ERP system client to avoid intruding into the data flow on the factory network.

Compatible Application-Layer Protocols

As mentioned earlier, Ethernet provides a platform used by higher-layer protocols to provide necessary communication services. To illustrate, Figure 1–12 shows the relationships between a number of different protocols.

From a services point of view, several are involved in reply/request exchanges invisible to the user, but are necessary to establish a communication session. These protocols include **DNS (Domain Name Service), RIP (Routing Information Protocol), SNMP (Simple Network Management Protocol),** and **ARP (Address Resolution Protocol).** From a structural point of view, IP and ARP use distinctive frames within the Ethernet frame format. Further description of these protocols can be found in Part Two.

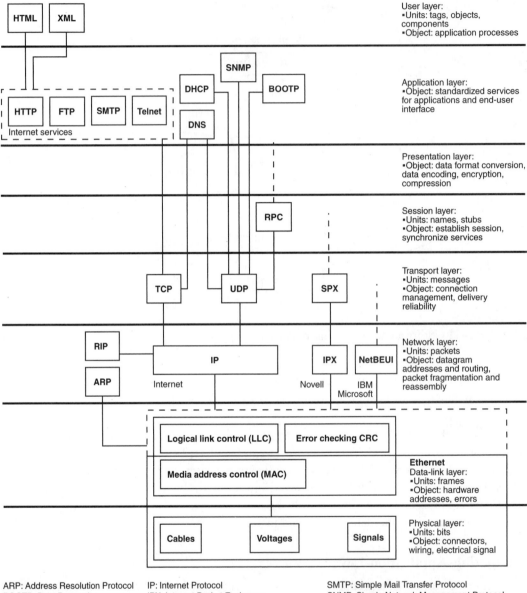

ARP: Address Resolution Protocol
BOOTP: Boot Protocol
CRC: Cylical Redundancy Check
DNS: Domain Name System
FTP: File Transfer Protocol
HTML: HyperText Markup Language
HTTP: HyperText Transfer Protocol

IP: Internet Protocol
IPX: Internet Packet Exchange
LLC: Logical Link Control
MAC: Media Access Control
NetBEUI: NetBIOS Enhanced User Interface
RIP: Routing Information Protocol
RPC: Remote Procedure Call

SMTP: Simple Mail Transfer Protocol
SNMP: Simple Network Management Protocol
SPX: Sequenced Package Exchange
TCP: Transmission Control Protocol
Telnet: Telecommunications Network Protocol
UDP: User Datagram Protocol
XML: Extensible Markup Language

FIGURE 1–12   Relationships between common Ethernet-based protocols

TCP/IP itself also accommodates higher-layer application protocols. We have already looked at HTTP; other familiar application layer Internet protocols include:

- *FTP (File Transfer Protocol).* Allows the user directly to transfer files between hosts.
- *SMTP (Simple Mail Transfer Protocol).* Supports basic e-mail delivery services.
- *Telnet:* Allows the user's device to operate as a terminal to run a remote host.

Compatible Network- and Transport-Layer Protocols

So far, this discussion has focused on TCP/IP. But Ethernet doesn't discriminate. It is an equal-opportunity employer. Alternatives to TCP and IP include:

- *IPX/SPX (Internetwork Packet eXchange/Sequenced Packet eXchange).* A Novell developed stack, IPX handles network addressing and routing; SPX handles packet sequencing.
- *UDP (User Datagram Protocol).* A TCP alternative that does not divide a message into packets to be reassembled later, but instead relies on the application to ensure the entire message arrives. UDP is typically considered as part of the TCP/IP family of protocols.
- *NetBIOS.* An IBM networking protocol for network and transport functions.
- *NetBEUI (NetBios Extended User Interface).* Microsoft's simplified implementation of NetBIOS for the Windows OS.

Incompatible Fieldbus Protocols

These and other Ethernet-compatible upper-layer protocols were developed in the 1980s. During this time, no protocols were introduced specifically for iE networks. Instead, automation and control companies were developing proprietary fieldbuses. Early fieldbus developments include:

- *Modbus.* Introduced in 1978 by Modicon (now part of Groupe Schneider) as a serial communications protocol originally operating at 9600 b/s.
- *ISA SP50.* A 1985 project of the Instrument Society of America (ISA) to define a serial bus to connect smart devices to replace the analog standard.
- *Profibus.* Begun in 1986 by a German consortium and eventually issued in 1990 based on an SP50 data-link layer and non-Ethernet physical layer.
- *Fieldbus Foundation.* The consolidation of 1986 French efforts toward a Factory Information Protocol (FIP), later taken over by WorldFIP and finally merged into Fieldbus Foundation in 1994. This standard provides digital, two-way serial communication for devices and for a distributed control network.

- *Controller Area Network (CAN).* Initiated in the 1980s by the German firm, Robert Bosch GmbH, with the support of Intel, initially for car-wiring networks and extending to low-speed fieldbus network for sensors and actuators.
- *DeviceNet.* Introduced by Allen-Bradley in 1994 and now an open standard administered by the Open DeviceNet Vendors Association (ODVA), it is based on CAN.

Today, there are over 200 fieldbuses. As originally developed, none run over Ethernet.

A USER'S PERSPECTIVE

Quite literally, Ethernet, TCP/IP, and HTTP protocols were developed by and for white-collar professionals in offices and classrooms; industrial networks and fieldbus protocols were developed by engineers for the factory. Consequently, bringing Ethernet technology to the factory floor involves a clash of cultures, performance expectations, and installation environments. It has taken nearly thirty years for these different worlds to converge, due to the following factors.

Culture Clash between Business and Industrial Networks

Ethernet—and the Internet—was indirectly inspired by U.S. Department of Defense-sponsored research into alternatives to the telephone network. Actual development was done by a loose collaboration of university researchers, professional standards organizations, and open software groups for *free.* In contrast, fieldbus was inspired by the necessity to improve factory productivity by using digital communications as an alternative to existing analog signals. The developers were automation and control companies with millions of dollars invested in developing protocols and related hardware and software.

Different Performance Expectations

Ethernet excels at delivering bursts of noncritical, kilobyte- and megabyte-sized packets for file-transmissions and messaging. Fieldbuses handle critical, instantaneous, byte-sized data streams from instruments.

Different Installation Environments

Ethernet is at home with PCs in electrically quiet offices. Industrial networks must withstand high electromagnetic interference (EMI) and provide their own electrical power on the wire.

Reasons for Convergence on Ethernet

By the mid-1990s, developments in information systems, business-application, factory-automation, and networking technologies provided compelling reasons for these worlds to unite.

Component	Proprietary cost	Ethernet cost
Gateway	$5,000	Under $500
Interface card	$1,500	$10
Cabling	$1.00/foot	$.06 to .08/foot

FIGURE 1–13   Comparison of costs between proprietary industrial-network and Ethernet-network components

ECONOMY

Figure 1–13 shows how mass production has made Ethernet components economically competitive with proprietary fieldbus hardware and software.

While low price is nice, reliability is more important to industry. Advances in Ethernet technology have met this concern also, as will be shown in later chapters.

SPEED

The capacity of the fastest fieldbus is 2 Mb/s. Ethernet handles 100 Mb/s and 1 Gb/s—and is moving toward 10 Gb/s.

DISTANCE

A single optical fiber can be run through a plant for over a mile without worrying about signal degradation or electromagnetic interference.

EXECUTIVE INSIGHT

New decision-making and management applications, such as **enterprise resource planning (ERP),** as well as **supply chain management (SCM)** and other collaborative solutions, presume the existence of high-capacity networks. These systems provide valuable decision-making support when they can quickly access accurate data from the factory floor—preferably without a lot of translation between protocols.

Today, the number 98 may be stored in a database and then pulled into a factory data analysis application, an ERP application's **cockpit,** or a Web-portal application's **executive dashboard** using an ordinary browser (Figure 1–9). These tools give a picture of what is actually happening in the facility, so decision-makers—including plant managers, suppliers and customers—can conveniently determine the best use of time, money, and manpower.

Also, the power of TCP/IP and http and Java-based applets is being embedded directly into **data acquisition (DAQ)** devices on the plant floor. The DAQ system shown in Figure 1–14 not only serves up Web pages, but also supplies data in a number of file formats that can be automatically uploaded into high-level factory automation applications. From there, the data can be pulled into manufacturing and enterprise information applications as required. If there are any problems or questions, any authorized person can take a direct look at the data in the DAQ right from their browser.

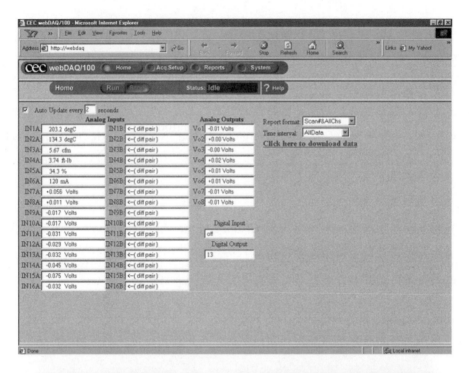

FIGURE 1–14 DAQ System capable of serving data in a Web page
(Courtesy CEC)

FUTURE PROOFING

In yesterday's business universe, information ran over Ethernet, and factory data—at least at the device and control levels—ran over factory networks. Now collaborative business applications employing data from the factory floor, plus Ethernet's growing capabilities, are pushing industrial users and vendors toward Ethernet.

The first industrial network protocol to run over Ethernet was introduced in 1999 (Figure 1–15). Major vendors followed suit with several introductions in the year 2000, effectively launching the Industrial Ethernet Age.

Taking advantage of Ethernet's flexibility, fieldbus organizations introduced **Modbus/ TCP, EtherNet/IP (DeviceNet for Ethernet), ProfiNet (Profibus for Ethernet),** and **Fieldbus Foundation HSE (High Speed Ethernet).**

These protocols make it possible, if not always advisable, for sensors, discrete and analog I/Os, and smart devices to communicate with controllers and supervisory devices over an Ethernet network.

In many respects, it took a shotgun marriage to bring Ethernet and fieldbus protocols together—the lure being overwhelming business benefits. The matchmaking took over a decade, with the evolutionary descent of Ethernet to the factory occurring in three stages (Figure 1–16).

But by 2001, the attractiveness of universal data integration and access, especially over the Internet, has dimmed. The "dotcom" crash, network outages, security holes, failed ERP implementations—have put a damper on e-business "solutions" that don't pay their way.

Nevertheless, the commitment to iE as a core network infrastructure continues. According to the ARC Advisory Group, the industrial market for Ethernet devices will grow to well over 4,000,000 nodes by 2005. According to other observers, it is already among the top three fieldbuses in factory networks.

As an open standard, Ethernet technology provides a common information infrastructure that has the flexibility, capacity, and economy to meet a wider range of needs. Consequently, different parties—even competitors—are relying on its capabilities, thereby making the technology more valuable with each passing year. These parties involve:

- Members of the Unix community.
- Hardware and software giants such as Microsoft, Sun Microsystems, Cisco, and Oracle.
- Manufacturing solution providers.
- Plant and automation control vendors.
- Fieldbus organizations.
- Application developers.
- Systems integrators.
- Control engineers.

Year	Ethernet/Internet	Business and information systems	Factory automation and control and fieldbus
1957	Advanced Research Projects Agency (ARPA) formed by Department of Defense		
1964	Paul Baran at RAND think tank theorizes a packet-switched network with no single outage point	MRP begun by IBM for J.I. Case (tractor maker)	4–20 mA analog signal introduced for instrumentation
1966	Robert Taylor and Lawrence Roberts connect ARPA-related universities with "packet switching" network		
1968			Dick Morley develops PLC (Programmable Logic Controller) to replace relay racks; GM orders first production line robots
1969	A four-node "ARPANET" connects three California universities and the University of Utah using 50 Kb/s phone line		
1970			GM uses Machine Vision
1972		SAP (Systemanalyze und Programmen-twicklung) founded	
1973	Robert Metcalfe's Harvard PhD thesis introduces Ethernet; E-mail comprises 75% of ARPANET traffic	Xerox develops Smalltalk object-oriented programming language Microsoft founded	
1976	Ethernet at Xerox runs at 2.94 Mb/s	Apple][computer debuts	
1977		Oracle founded	
1978	Early OSI model published	VisiCalc spreadsheet Baan founded	Modbus introduced by Modicon

FIGURE 1–15 Developments in industrial and Ethernet networking

Year	Ethernet/Internet	Business and information systems	Factory automation and control and fieldbus
1979	*Digital Equipment Corporation, Intel, and Xerox create DIX Ethernet standard for 10 Mb/s*		
1980		MRP extended to MRPII by JD Edwards	Allen-Bradley introduces Remote I/O
1981		IBM PC	
1982	TCP/IP becomes ARPANET standard, begins "internet"		
1983	*Institute of Electrical and Electronic Engineers (IEEE) releases an Ethernet standard (802.3) interoperable with DIX*		
1984	Domain Name System introduced	Cisco Systems founded by Sandy Learner and Leonard Bosack Macintosh introduced	
1985			ISA SP50 begins to define serial bus
1986	National Science Foundation creates backbone (NSFNET); Internet Engineering Task Force (IETF) is formed	PeopleSoft founded	Profibus development begins
1988	NSFNET upgraded to T1		
1990	Over 100,000 hosts on Internet	Baan rolls out ERP; Bill Joy at Sun proposes object-oriented programming environment	
1991	Tim Berners-Lee introduces World Wide Web	Sun develops Java	
1992		Windows 3.1	
1993	WWW presented at Online Publishing 93; White House goes on-line		

FIGURE 1–15 Continued

Year	Ethernet/Internet	Business and information systems	Factory automation and control and fieldbus
1994	Marc Andressen launches Netscape (Mosaic Corp.)	Oracle database released for PC; Microsoft memo notes Internet	Allen-Bradley introduces DeviceNet
1995	WWW surpasses FTP data; first internet-only radio starts broadcasting	Apache Group organized	Foundation Fieldbus developed by ISA/IEC
1996	MCI upgrades Internet backbone to 622 Mbps; browser war between Netscape and Microsoft		
1997	*IEEE 802.3x bypasses collision detection to achieve "full-duplex" communication*		
1998	Network Solutions registers 2,000,000 domains		
1999	*IEEE 802.3ae for 10 Gb/s under development*	ERP becomes Internet enabled	Modicon introduces Modubus/TCP, the first fieldbus protocol to run over Ethernet
2000	Over 22,000,000 Web sites; "dot com" crash begins		Three other iE protocols introduced (Ethernet/IP, ProfiNet, HSE)
2001	Dense wavelength division multiplexing splits wavelength allowing 1.28 Tbps per fiber; *IEEE 802.3ah developing "last mile" Ethernet at 10 Mb/s and 1 Gb/s over optical fiber*	"Dot com" crash aggravates an eighteen-month manufacturing slump, causing a major re-evaluation of information technology, from the Internet to ERP systems	
2002	Overcapacity in long-haul fiber networks; metro networks begin build out as fiber becomes more accessible to businesses in major cities	Distributed networking and security concerns follow 9/11/01 fears of vulnerability	Major vendors include Ethernet ports in PLC and SCADA systems; Ethernet TCP/IP being embedded in I/O modules and some sensors

FIGURE 1–15 Continued

34

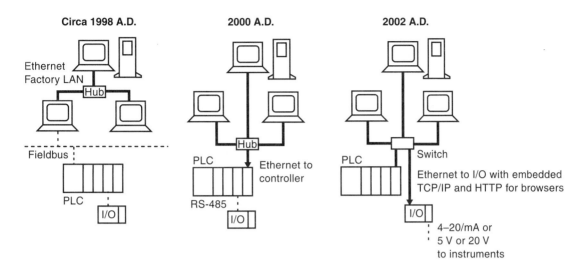

FIGURE 1–16   The evolutionary descent of Ethernet to the factory floor

As a sign of vitality, Ethernet technology continues to evolve. New IEEE Ethernet standards include wireless (802.11), 10 Gb/s fiber (802.3ae), and "first-mile" connectivity direct to workplace or home (802.3ah).

In a business and manufacturing environment that increasingly values flexible, economical, fast, and collaborative information access, the future of iE is wide open.

SUMMARY

- A business's enterprise network is comprised of all its LANs and WANs, but usually not its industrial networks.
- The demand for instant and universal access to data from the plant is ending the isolation of industrial networks.
- The OSI model is used as a conceptual reference to understand how Ethernet functions.
- The code inside a data packet can tell you all the network interactions involved with a piece of data, plus the data itself.
- Different protocols play a specific role at each stage of a request-and-reply communication exchange.
- The header contains protocol control information that allows the payload to be properly processed.
- The Ethernet protocol enables data encapsulation, TCP enables packet sequencing, and IP enables datagram fragmentation.
- The TCP/IP stack enables Internet communication.

- A switch reads network the address in an IP header and passes the packet to the proper network.
- A router reads the port number in a TCP header and passes the packet to the proper port.
- A gateway reads the application header and translates the data into a form suitable for a different type of physical network.
- XML-tagged can be displayed in browsers and human machine interface (HMI) panels and imported into data historians and databases.
- There are over 200 fieldbus protocols. As originally developed, none run over Ethernet.
- According to the ARC Advisory Group, the industrial market for Ethernet devices will grow to well over 4,000,000 nodes by 2005.

? REVIEW QUESTIONS

1. What application layer service uses TCP layer protocol port 80?
2. What function do sockets perform?
3. Name another protocol suite that can be used on Ethernet in place of TCP/IP.
4. What OSI layers do not need to be used for Internet communication?
5. What is a port?
6. What is IP?
7. What does the sequence number in a TCP header do?
8. What is fragmentation?
9. What protocol provides address resolution?
10. What is screen scraping?
11. How many years elapsed between the introduction of Ethernet and the first iE protocol?

Why Fieldbuses Only Go So Far: Factors in Networking the Factory Floor

2

OBJECTIVES

After reading this chapter, you will be able to:

- Understand how a network connects the components in a control system
- Appreciate the benefits and difficulties of using digital communication for control
- Identify three stages in the development of control networking, culminating in industrial Ethernet
- Understand why several industrial control protocols have adopted Ethernet as a networking platform

The previous chapter showed how Ethernet delivers a particular piece of data to a Web browser–based **dashboard.** The focus was on where the data packet was going, not where it came from.

Now it is time to get our hands dirty. This chapter presents the basics on industrial networking. To illustrate the discussion, we continue to follow the travels of the number 98. But this time the journey starts down at the **sensor level** on the factory floor, where the value 98 originates, and ends at the **plant level,** where the number can be pulled from a database and displayed on a Web page, as shown in the previous chapter. Along the way, readers unfamiliar with the factory environment may refer to the following list of terms on plant automation and control (Figure 2–1).

BASICS OF CONTROL COMMUNICATIONS

To focus on the large subject of networking for factory automation and control, it helps to concentrate on one word: data. Data flow is the fundamental reason for the existence of a control network. Data originates from components in the control loop. Control loops can use pneumatic and analog signals, precursors of modern digital communications discussed in the following section.

Analog signal	A fluctuating wave of electric current that represents by "analogy" the flux of a control variable, such as the rise and fall of temperature or pressure.
Closed-loop control	A system in which an output is fed back as an input to adjust the system's performance.
Controller	A stand-alone device, such as a PLC, or a program in a PC that receives outputs of variables, performs control calculations, and generates an output used to adjust the system.
Data acquisition system (DAQ)	An digital instrument or PC that gathers data from sensors. It is used in conjunction with amplifiers, multiplexers, and analog-to-digital converters. Often employed in process industries.
Digital signal	An electric signal that uses defined levels of changing current or voltage to encode binary values (0,1).
Devices	In control, field devices are the physical elements that affect a controlled system and are typically located at a distance from the controller. Input devices can be switches (on/off, pushbutton, limit, proximity, etc.); output devices can be lights, motor starters, pumps, and solenoids that drive valves. Measuring devices include flow, temperature and pressure sensors as well as more complex instrumentation.
Distributed computing, distributed control, distributed I/O	As a general concept, a distributed approach breaks a complex centralized system into modules that can perform operations at dispersed locations and transmits outputs to other devices. The distributed concept depends on inexpensive processors and pervasive networking. Thus, distributed computing involves modular program components (such as Java applets or CORBA objects) instead of a single program; distributed control involves control functions embedded in "smart" field devices instead of a single PLC or PC; distributed I/O involves I/O modules in the field instead of local I/O boards in a centralized PLC.

FIGURE 2–1 Plant automation and control terms

DCS (Distributed Control Systems)	Evolving from centralized process control computers in the 1960s, DCS handles continuous-flow processes involving open-loop and analog control. Due to the nature of continuous and complex batch-process applications, DCS requires "real-time" performance in both the control system and control network. Since these applications are associated with large chemical, water, waste-water, and utility plants, the terrain makes it advantageous for devices and controllers to be located near the operation and away from a centralized control room. A DCS is networking-intensive because the control information is usually transmitted to a central control room.
Embedded control	The incorporation of a microprocessor in instruments and sensors to perform control functions and network communications.
Fieldbus	A protocol for two-way digital communications between field devices and controllers. All control and networking devices must be compatible with a given fieldbus standard for data transmission to succeed. Fieldbus is an alternative to pneumatic (3–15 psi) and electronic analog (4–20 mA) communication.
Home run wiring	Wiring that runs from the local I/O of a centralized PLC to field devices. The precursor to fieldbus.
HMI (human-machine interface)/MMI (man-machine interface)	A display panel used with a machine controller, PLC, or a PC that allows the operator to see a graphic representation of the operation for monitoring and control.
I/O Module	Input/output is a general term for data entering or leaving a device. In a control system, an I/O module contains circuits to process a number of inputs received from various measuring devices into digital outputs that are transmitted to a PLC or PC. Digital modules interface with on/off sensors/actuators; analog modules use D/A and A/D converters to interface with analog instrumentation; intelligent I/O modules can provide an interface for high-speed counters, programming, motion control, or communication.

FIGURE 2–1   Continued

Instrument	A device that detects a state like a sensor, but can also observe, measure, and provide data, and may even control a value.
Intrinsic safety	A characteristic of control wiring or devices whereby voltages and currents are kept below the levels that can ignite an atmosphere.
Manufacturing execution systems (MES)	A high-level information system used in the plant for quality control, document management, plant-floor scheduling, and dispatching. An MES tracks work-in-process (WIP) by collecting data on product routing and tracking, labor, resources, and production. It also collects "live" information about setups, run times, throughput, and yields to identify bottlenecks and capacity problems. Consequently, the MES is critical to plant floor *and* enterprise communication by delivering real-time order status to the supply chain, by enabling an available-to-promise capability for the customer, and by updating ERP systems for business managers.
PLC	A controller with a microprocessor, memory, and programming to control large numbers of discrete elements using very fast I/O scan times. PLCs are used with other computers and applications, such as human-machine interface panels and data historians for data storage and evaluation.
Point of control/data point	A value representing the state of a controlled variable, such as on/off or a temperature reading, transmitted as bits, or a measurement, usually transmitted as bytes.
Repeater	A network device that enhances electrical signals so they can travel greater distances, allowing a larger number of nodes to be connected to the network.
Scan time	The time required for a PLC to completely execute a program once, including an update of I/O values.
SCADA (Supervisory Control and Data Acquisition)	Evolving from PLC operator interfaces, SCADA is a software-based system used for process monitoring and data acquisition, as well as control. Supervisory control uses software-based logic to perform the same functions as the PLC with a PC. Software can add production management capabilities to allow supervisory control systems to perform at an MES level.

FIGURE 2–1   Continued

Sensor/Transducer/Transmitter	A transducer is a device that converts signals from one physical form (air pressure) to another (electrical current). In a broad sense, a sensor is a transducer that uses the state of a physical phenomenon as an input and returns a quantitative measure as an output. A sensor functions as a transmitter when the measurement is transmitted as a digital signal to a PLC or PC.
"Smart" Device	A sensor with the embedded intelligence to identify itself on the network.
Terminator	A device used to absorb the signal at the end of a fieldbus cable.
Workstation	An area where a single operator monitors and responds to one or more conditions. It may use an HMI or PC with integrated messaging and graphics.

FIGURE 2–1   Continued

Data Flow

Data must flow within the control system to keep the operation running properly, moment by moment. But it must also flow beyond the control system where it can be:

- Recorded to verify the status of the operation.
- Used in plant management and business-decision applications to improve all phases of the operation.
- Sent to supply chain partners and collaborators outside the business to notify them of relevant status, quality, and inventory issues.

As applications for data and data-intensive decision-making applications have proliferated, networks are asked to supply more data, faster. The goal: up-to-the-second control data for the operator on the factory floor—and up-to-the-minute operational data for the manager on the top floor.

This chapter is about the quest for a control networking technology that satisfies both objectives. The ideal industrial network combines the high capacity and flexibility of Ethernet with the reliability and performance of early control networks. To understand this dual mission, it helps to understand the communication needs of a basic control loop, then examine how networking technology is used for communication between multiple devices, controllers, and applications.

Control-Loop Components

CONTROL SYSTEMS

Control systems are employed in industry to adjust, monitor, and record the condition or status of machinery, products, and processes. At specific times or events, the operation is measured at different points. The data are then used to make adjustments that improve the condition or status. This communication cycle of measuring, adjusting, and improving the system is known as a **control loop.**

In a basic control loop, the control network connects four components (Figure 2–2):

1. A **sensor** or instrument used for measurement.
2. A **controller** for calculating both deviances between the measurement and the setpoint and the degree of adjustment necessary.

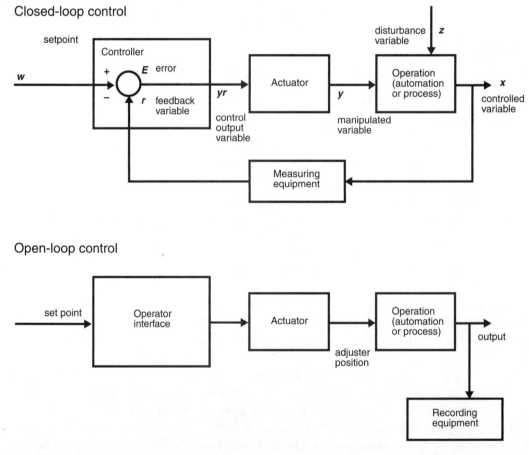

FIGURE 2–2   Signals and elements in closed- and open-loop control

3. An **actuator** for making the adjustment.
4. A **gauge** or panel for viewing the state of the operation.

CLOSED- AND OPEN-LOOP CONTROL

How components communicate with each other depends on whether they are used in closed- or open-loop control. In **closed-loop control,** the controller receives measurements in the form of numerical inputs. Then the controller generates outputs according to programmed parameters to drive the actuator. Since the outputs are used to make changes in the operation, which is then remeasured to make further adjustments if necessary, a feedback loop is created. A feedback loop is the hallmark of closed-loop communication.

In an **open-loop control** system, outputs are usually monitored and recorded; there is no feedback loop. The open-loop method is often used in process industries, such as water treatment, where monitoring, alarming, and data recording are the primary functions.

At the most basic level, individual wiring can be used to connect all the devices directly to the controller. With multiple devices or controllers, a control network can be used to transmit data from multiple sources over a common wire.

The control network transmits inputs and outputs (I/O) in the form of **continuous values** or **discrete values.** Continuous values take the form of an unbroken range of numbers, such as fluctuating measurements of voltages, pressures, flow rates, temperatures, and position coordinates. Discrete values are represented by numbers within a finite set, such as a 0 or 1 indicating that a switch (i.e., a pressure, flow, or position switch) is off or on.

The network is also used to transmit instructions to the controller, such as algorithms, set point values, batch recipes, and recipes of sequential steps.

Pneumatic Communication

In early twentieth-century industry—when the telephone was still a novelty—control networks were pneumatic, not electric. Compressed air was used to convey signals between the measuring device, gauge, and actuator.

Since a pneumatic system clearly illustrates the various aspects of networking in any control system—whether pneumatic, hydraulic, or electric—it is worth exploring this early stage of control to understand the workings of industrial networking in general.

Of course, in a pneumatic control system, data is not sent over copper wire; rubber tubes are run from an air compressor to the measuring instrument (Figure 2–3). Then each tube is connected to a gauge and an actuator. A stream of air pressurized to 15 pounds per square inch (psi) fills the tubing. Changes in pressure are used to indicate changes in the variable being measured.

To control a fluid level (the **variable**), a valve-and-float device (the **sensor-transmitter**) is set up on a tank. As the float sinks, it closes a valve to restrict air flow in the tube connected to the gauge. A low float lowers the air pressure to signal a low fluid level; a high float increases air pressure to indicate a high level. As the level changes, the pressure rises or falls on a sliding scale of psi, which is the characteristic of an **analog** signal.

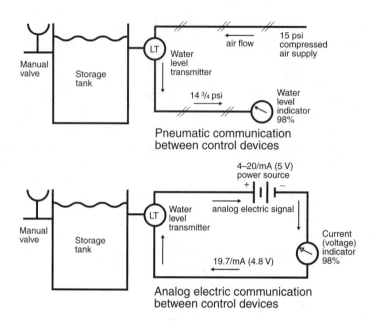

FIGURE 2–3   Simple pneumatic and analog electric control communication

FUNDAMENTAL FEATURES

Although this "early" form of control communication is a simple example, it is still employed today in hazardous applications where electrical wiring cannot be used. It exhibits four features that characterize any successful industrial-networking technology:

1. *Simplicity.* Problems in the network can be easily traced to one of two variables: a puncture in the tubing or a problem with the air compressor.
2. *Economy.* Rubber tubing is an inexpensive material to transmit signals.
3. *Safety.* In flammable applications, pressurized air won't generate an electrical spark; in chemically reactive applications, rubber tubing won't oxidize.
4. *Flexibility.* Physical flexibility allows tubing to be strung throughout the plant; application flexibility means air pressure can operate either a gauge (the instrument) or a valve to regulate flow (the controller/actuator).

In this rudimentary example, it is clear that pneumatic technology can succeed in physically connecting all the components in a basic control system—sensor, gauge, actuator, and controller.

PERFORMANCE REQUIREMENTS

In addition to meeting physical requirements, a control network must also handle several performance issues:

Signal accuracy and resolution.   A control network must accommodate signals that accurately represent the underlying process or operation. An analog signal has the advantage of itself being a continuous **sine wave** that is an imitation *analog* of the continuous flux in the process.

In the pneumatic example, a gauge can be calibrated to read "100%" at 15 psi and "0%" at 3 psi (calibrating "0%" with live pressure in the line provides a reference point to distinguish a **dead system** with 0 psi). As the tank level moves up and down, so does the air pressure. A tank that is two-thirds full cuts the air pressure to 8 psi, so the gauge needle simply swings to the 8 o'clock position without the rounding error of digital numbers. The simplicity of analog measurement is a drawback, however, when data needs to be communicated with digital precision—for example, when the air pressure hits 14.76 psi, it may be difficult to read the needle to see that the tank is precisely 98.0% full.

Signal integrity.   A control network must keep data signals from degrading during transmission. To avoid **dirty signals** that throw off gauges and valves, pneumatic devices must be supplied with clean, dry air. Water in the line will throw off the pressure and the accuracy of the readings. In remote applications, the length of tubing can only run so far before dropping air pressure requires another compressor (the **repeater**) to regenerate the signal.

Data recording and accessibility.   A control network must be able to handle the required quantity and quality of data. With one variable (fluid level) at one control point (the tank float), a simple pneumatic system has no problem handling the data. But when measuring and controlling many variables, adding tubes becomes physically awkward. A control network must be easy to expand. It must also easily interface with monitoring and recording devices. Data management is a problem for our simple pneumatic system. Unless the measurement is written down, the value displayed by the gauge is literally gone with the wind.

Analog Electric Communication

The telegraph and telephone proved that thin copper wire could transmit signals. Over the years, various schemes used electrical signals for control communication. Eventually, just as the 3–15 psi became an accepted pneumatic standard, a 4–20 mA current became an electric signal standard in the 1960s, and it is still used in analog instrumentation today.

Electric analog communication is simple and economical. Plus, wiring is very flexible when running long distances and connecting multiple control points. But with electricity, safety is a problem. The current has to be limited to avoid a spark. Signal integrity is also an issue. Wires need to be shielded to prevent **electromagnetic interference (EMI)** from affecting the output signal's characteristics.

A benefit of electric analog communication is improved data recording. When using an analog electric signal for measurement, a resistor can be applied to the 4–20 mA current flow to convert it into a 1–5 V signal. Thus, either a 19.68 mA or 4.84 V analog signal indicates a value of 98%. The current can also drive amplifiers for small motors attached to ink pens (known as **pen plotters**) to record the data on a roll or spiral sheet of paper—an advance over manual data entry.

MODERN DIGITAL COMMUNICATION

The benefits of pneumatic and analog electrical control look trivial today, but that is only in comparison to their successor—digital communications. Digital communication is ideal for transmitting on and off discrete valves. It has a more difficult time handling continuous analog valves. Nonetheless, it sets the standard for industrial networking.

Digital Signals

For a digital signal to indicate an increase or decrease in voltage or current, it must be encoded so a drop in voltage or current indicates 0, and a rise indicates 1.

For example, a low voltage level between 0 V and 0.8 V can be used to represent a 0, and a high voltage level between 2.0 V and 5.0 V can represent 1. (The range between 0.8 V and 2.0 V is unused.)

These high and low voltage levels can be sequentially arranged as a binary bit stream. The bits are then transmitted between digital devices that decode the stream. One method of encoding is **Manchester encoding.** Each bit of data is transmitted within a defined period of time **(bit period),** which is defined by a **clock.** For a binary 1, the first half of the period is high, and the second half is low. For a binary 0, the first half is low, and the second half is high. Because Manchester encoding renders the signal pulses into meaningful bits, the bits can then be grouped into bytes according to ASCII encoding, as described in the previous chapter.

As shown in Figure 2–4, the hex values for the number 98 can be transmitted in the form of a digital signal shape that corresponds to the ASCII binary values for 9 (00111001) and for 8 (00111000).

Digital Complications

Besides encodability, digital signaling offer other benefits. Signal integrity is easily engineered, because the wave shape can be designed to withstand high levels of EMI. Data recording and accessibility is much easier, because the digital signal is already in the binary language used by computers.

RESOLUTION

Unfortunately when it comes to signal accuracy and resolution, digital communication presents complications in certain control applications.

To understand why, consider how analog signals and digital signals are produced. For example, a curve can be easily drawn with a pencil and compass in one sweeping motion. In the same way, an analog signal can easily reproduce the natural curves of a measured phenomenon, because an analog sine wave is already a good *analogy* of a curve.

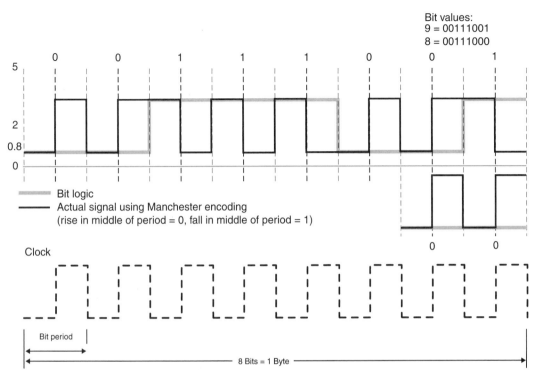

FIGURE 2–4   Digital signal of ASCII binary values for 9 and 8

Alternatively a ruler can be used to measure a few points on the curve; then dotted pencil marks can create a crude approximation of the arc. In the same way, digital signals reproduce a phenomenon by putting together pieces of information. But to get the resolution necessary to get a smooth curve requires sampling thousands of points to recreate the slope of the curve.

The problem of resolution is different in discrete control. A tank may simply need to be refilled. In this case, a singular event—such as opening a valve to fill the tank—can be indicated by binary values: 0 equals *close;* 1 equals *open.* The only issue is taking samples fast enough to determine status in the desired time frame.

With binary inputs from relays, switches, counters, and digital transmitters, there is no need to depict a sweeping curve. The control system simply needs to process—and the control network needs to transmit—bits of data at regular time intervals fast enough to characterize events as they occur.

When digital signals are used to describe continuous phenomenon, complications arise. In terms of our illustration, the arc must be converted into point values. With analog instrumentation, the state being measured is naturally represented as a sine wave (Figure 2–5).

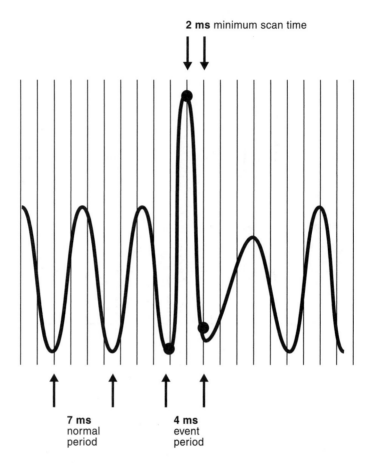

FIGURE 2–5   Sine wave and scan frequency related to event

SAMPLING

To be converted to digital, the analog signal must be measured **(sampled)** at regular time intervals and then be processed by an analog-to-digital converter (A/DC) (Figure 2–6). To get a good representation of the wave, values for both the positive and negative peaks must be captured. Consequently, the sampling frequency must be at least two times faster than the fastest frequency in the signal—a principle discovered by **Harry Nyquist,** a Bell System engineer who pioneered digital encoding to speed up the telegraph.

If the wave from the process is not sampled fast enough, differences occur between the shape of the analog signal and its digital reconstruction. This error is known as **aliasing.**

A signal is also misrepresented when the environment and the hardware adds noise and error during the conversion process. To eliminate these problems, digital instruments and analog/digital devices need to sample at speeds many times faster than the highest frequency component of the signal being sampled.

Therefore in a closed-loop control system, network response time is critical. The minimum speed of the system must be the speed of the closed loop, plus the feedback loop. This speed must be maintained to keep the system stable, which is referred to as a **steady-state.**

Depending on the method of calculating the sampling rate, the controller may need inputs many times faster than the steady-state bandwidth. In an open-loop control system, network speed is also critical, but more so bandwidth, as explained below.

More Complications: Real-Time, Deterministic, and Synchronized Performance

As just mentioned, the combined speed of the input loop and feedback loop must keep up with the speed of the system to maintain stability. Because an entire system is involved, the speed of the CPU, software, and network are all factors affecting real-time response.

Unlike an office LAN, an industrial network cannot be clogged by large file transmissions. Otherwise, critical control or alarm data will arrive too late. Industrial networks are designed to keep **worst case** transmission delays within tolerances of milliseconds. (Office LANs will sacrifice guaranteed response time to avoid the expense of adding bandwidth or creating separate collision domains.)

Within milliseconds, the PLC polls the device to acquire data on the event, receives the data from the I/O module on the sensor bus, makes calculations based on the appropriate control algorithm (P/I, PID, etc.), then transmits the necessary updates. Inputs may be derived from master-slave polling or client-server/publish-subscribe broadcasts, as explained later.

SCAN TIME

With real-time response, the **scan time**—the time to execute the entire PLC operation and communication cycle, including the I/O update—is set fast enough to handle each and every event *at a speed based on the worst-case scenario* (see Figure 2–5). This can be accomplished by setting a fixed polling time frame and scheduling all subsequent tasks to fit within it.

Because the timing is calculated on a predictable basis, the response is said to be **hard real-time.** And because the timing is able to capture each and every event as they occur, the performance is said to be **deterministic.**

The benefit of real-time response and deterministic performance is that the control system is adjusting variables to actual events, not making a late adjustment to an event after the fact. Obviously, there are severe consequences if the timing and logic are out of synchronization. The control system is forced to play a losing game of catch up as late adjustments to events already spiraling out of control make matters worse.

Transmitting the continuous flow of real-time data may require synchronized messaging between devices—a requirement in axis coordination of drives and in transmitting live voice and video streams. In these cases, the generated information must be received at a synchronized fixed rate to prevent data overflow and underflow. To create a synchronized system, a *master clock* is employed on the network to allow sending and receiving devices to establishing timing references. Both the source and destination devices use the common clock to keep pace.

BANDWIDTH DEMANDS

The demands of real-time response directly impact network bandwidth requirements. For example, consider a system with 125 discrete and 25 analog I/O updates for each of four PLCs (totaling 500 discrete and 100 analog channels) and a required 25 ms polling rate. Since 8 discrete I/O bits comprise a single byte, and each analog value takes 2 bytes, multiplied by 20 to ensure adequate sampling for closed-loop bandwidth, then 1,500 bytes of data are involved in each system update. If system overhead—such as error checking and addressing—comprises 30% of the network traffic, then the network must be able to handle 2,000 bytes per update cycle. That multiplies out to 80,000 bytes per second times two for each analog value, which requires a network bandwidth of 1.6 Mb/s. Operations that require a faster polling rate, such as motion control requiring scans under 5 ms, can be served by a dedicated network segment to handle that traffic.

Other factors—such as amount of trending and reporting traffic, A/D conversion, and the number of alarm points—may add to network traffic, change the bandwidth calculation, and may ultimately require a switch to isolate a network segment in order to maintain real-time response within the allocated bandwidth.

In industrial process applications, networks carry the heaviest load with SCADA packages that use set polling times (scans) to acquire data on status. Data packets must arrive within the allotted time; if timing gaps occur, the system may assume an error condition. The scan times may be very short to ensure data is collected when alarms or events are generated. Updates are constantly sent to the SCADA system database. Where appropriate, the traffic and timing burden can be relieved by using an application that **reports data by exception,** that is, only when there is a change in status.

Digital Bottom Line: Amazing Data Accessibility

If digital communication presents such complications, why employ it for control? There are three benefits—all related to the flow of information:

1. *Speed.* Due to multiplexing and other technical reasons, digital signals can transmit more data much faster than an analog signal.
2. *Data storage.* A digital signal is easily processed by a computer and stored on permanent media, such as magnetic tape and hard disks.
3. *Flexibility.* Digital information can be transmitted between many different devices. When data is shared in a useful, accessible, and logical form, decision makers benefit by getting more accurate and timely information.

Digital communication requires more complex devices than an analog system (see Figure 2–6). The **transducer** must be able to output an encoded signal that can be processed by a compatible controller and actuator. And the display can no longer be a simple analog gauge. What is required is a device—such as an HMI panel or a PC—that can decode the digital signal and display it on a monitor in alphabetical/numerical form.

Components in analog system

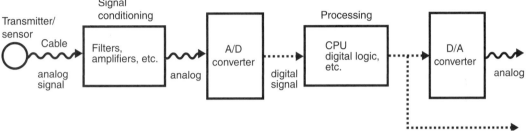

Components in digital system

FIGURE 2–6   Comparison of components in analog and digital control communication

Digital Standards for Low-Level Communication

Today, many digital communication standards are available to transmit bit-sized data from sensors and instruments. HART and RS-232 are two popular choices.

HART (HIGHWAY ADDRESSABLE REMOTE TRANSDUCER)
This digital communication protocol operates on top of a continuous analog 4–20 mA signal. It was introduced in the mid-1980s by Rosemount, which handed it over to the non-profit HART Communication Foundation. Because this method is a clever bridge between analog and digital communication, HART is the most widely used protocol in analog-oriented process industries. The HART standard is used to communicate within the control loop, for example, between a measurement device (transmitter) and a valve positioner. Communication is accomplished by changing (**modulating**) the frequency of the analog signal's sine wave.

As shown in Figure 2–7, HART uses an encoding scheme where a slow 1200 Hz (1200 sine waves per second) signal conveys a high bit value of 1; the fast 2400 Hz (2400 sine waves per second) signal conveys a low bit value of 0. The amount of data communicated at this rate is limited by the slowest rate (1200 Hz), which amounts to only 1200 bits per second when using 1-bit-per-symbol encoding. This transmission rate makes HART suitable as a **sensor bus,** that is, for low-level sensor data traffic, such as 1-bit values for on/off (contact) I/O, simple transducers, and measurements.

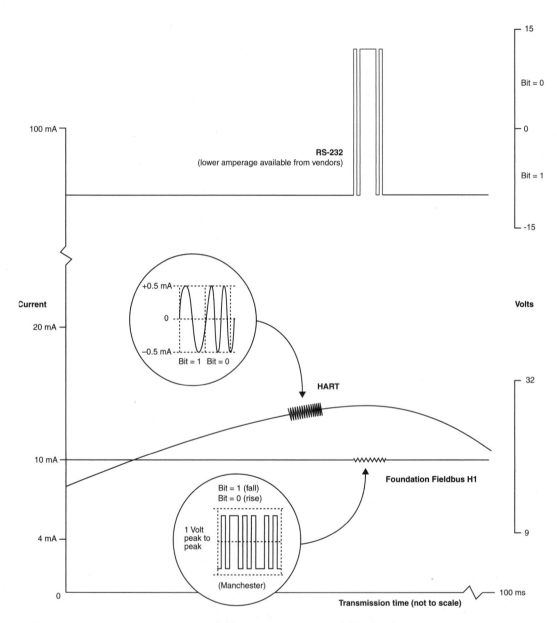

FIGURE 2–7   Differences in RS-232, HART, and Foundation Fieldbus H1 signaling

RS-232

This well-known serial-communication standard was the **recommended standard** (hence, RS) issued by the Electronic Industries Alliance (EIA) in 1969. EIA is also responsible for RS-422, RS-485, and RS-423. In serial communication, bits are transmitted sequentially, requiring separation by techniques of time division.

For bit transmission, RS-232 employs **voltage mode** signaling, using a low current and bipolar voltage, which means voltages swing between positive and negative. For example, the negative range of –3 V to –12 V is encoded as a bit value of 1, the positive +3 V to +12 V, a bit value of 0 (which differs from most other signaling standards where the **positive logic** value is marked as 1). Unlike HART, RS-232 is not a protocol, because it only defines signal levels, connector pin outs, and cable characteristics.

While the transmission rate of RS-232 is faster than HART, it is still only appropriate for connecting sensor-level devices. The signal is easily degraded by electromagnetic interference, so the wiring length is limited for use with devices in a localized control loop.

FIELDBUS COMMUNICATION

While useful for low-level, bit-sized, digital communication, HART and RS-232 cannot handle high-volume, byte-sized data traffic. Nor are they sophisticated enough to connect multiple devices. They are applied in a small, point-to-point network where the control loop does not extend beyond a local PLC, HMI panel, or PC.

As the digital revolution picked up speed in the 1980s, digital control devices were introduced that incorporated new communication protocols known as **fieldbuses.**

The term *fieldbus* applies to a large family of two-way, digital communication protocols that were specially developed to overcome the physical and performance limitations of low-level digital and analog standards.

A full fieldbus protocol can handle byte-size data for complex transmitters and valves, plus diagnostics or control information. The capacity to handle larger data blocks is needed to transmit multiple variables, plus device management information. Any control device requiring extensive communication for configuration requires a full fieldbus.

The Evolution of Fieldbus

The advent of full fieldbus protocols opened up—quite literally—a whole new level of interconnection in control and factory networks, as well as advances in devices that could speak the language of fieldbus. As shown in Figure 2–8, these developments occurred in several stages.

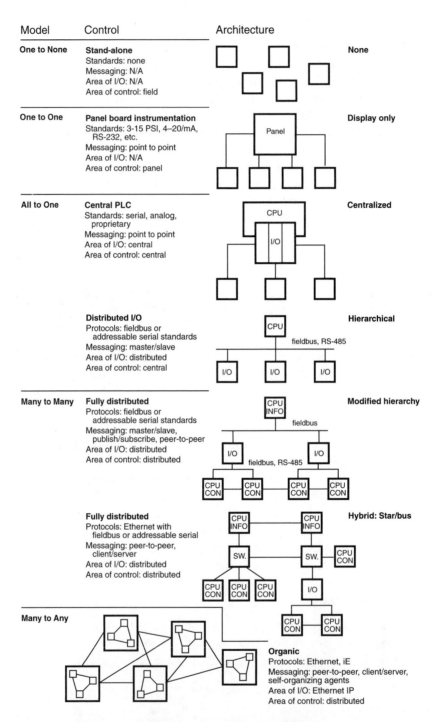

FIGURE 2–8   Models and corresponding control network architectures

ONE-TO-ONE PHASE (PRE-FIELDBUS ERA)

The one-to-one level of interconnection uses pneumatic, hydraulic, and electromechanical relays. Relays are wired arrays that serve as switching and timing mechanisms for input devices, such as push buttons, scales, and photoelectric eyes. Because each input and output is individually wired according to its own control logic, relays are said to provide a **one-to-one I/O ratio.** While electromechanical relays can be used in specialized applications, the cost of wiring and maintenance led to their general replacement. Today, transistors on circuit boards are used to re-create relay logic electronically.

ALL-TO-ONE PHASE (FIELDBUS ERA)

The beginning of the all-to-one phase can be traced to the 1970s when the PLC became popular. A single PLC can be programmed in relay ladder logic to handle a number of inputs and outputs. A PLC takes up far less space than electromechanical relays or circuit board components and is easier to program. Since PLCs were initially expensive, the network was build around a single, centrally located PLC.

Home run wiring is used to connect field devices to I/O blocks inside the PLC cabinet. But modifying or expanding a network comprised of miles of wires is a hassle. To overcome this difficulty, various networking solutions were introduced, collectively known as *fieldbus.* In a fieldbus network, the controller is not directly wired into the device. Rather the device is connected to a bus network. The CPU acts as a **master device,** which controls the bus and transmits messages to field devices, known as **slaves.** Slave devices cannot directly communicate with each other.

With a bus topology, all devices at a given level are connected sequentially on one long cable. Depending on the type of network, many nodes can tap into the bus and may or may not communicate with other nodes on the same cable. Another tap connects a drop cable to a PLC on a higher-level bus. A terminator is required at each end of the cable to match the impedance and to absorb signals to prevent reflections from creating noise. The disadvantage is that if a cable breaks, communication between nodes on the affected bus are cut.

Advances in PLCs, microprocessor-based workstations, and PCs and the use of master-slave communications embedded in field devices allow I/O boards to move out of the cabinet and into modules in the field. The ability to receive digital data from many I/O devices over one bus ushered in the era of **distributed control systems (DCS).** The development of DCS was particularly advantageous for process industries, which has the problem of managing a widespread network.

The use of the term *distributed,* however, can be misleading, since traditional DCS equipment is housed in a centralized control room. In the all-to-one phase, I/O data still go to a central computer or control system. The central processing unit (CPU) scans every device and process inputs and outputs for each control loop. Consequently with a large number of devices, a powerful and expensive CPU is required to handle analog-to-digital/digital-to-analog conversion and algorithm processing.

In the all-to-one phase, a fieldbus network is suitable not only for process control, but also for factory automation. Factory automation involves discrete control of production machinery, such as in automobile assembly and electronic component production. Therefore, the network must accommodate fast, short transmissions to ensure real-time response. To this end, the number of devices, length of the bus, and the size of data packets are minimized with a fieldbus. A simple network may involve 30 to 50 devices, a 100-yard bus length, and 1 ms to 5 ms transmission times.

Process control involves continuous operations, such as chemicals, paints, and power production. In process control, the network needs to operate in hazardous environments and over long distances and handle transmissions from many field devices. A typical application can involve hundreds of devices, miles of bus cabling, with 10 ms to 15 ms transmission times.

A full fieldbus suite—such as DeviceNet and ControlNet—fits comfortably in a **three-tier architecture**—a bus connects sensors and actuators at the bottom level; another bus connects intelligent devices and controllers at the mid level; and yet another connects PLCs, DCS, SCADA, and PCs at the high control and supervisory level.

Many-to-Many Phase (Open-Fieldbus/Ethernet Era)

By the early 1990s, PLCs and microprocessors gained in power and decreased in price. The trend is to move away from a hierarchical model, because every layer is a potential bottleneck. Microprocessors in "smart" sensors and devices perform basic control and communication functions in the field. The centralized CPU is used to run information applications—such as MES—while control is accomplished via the network.

In some cases, the intelligent sensor-transmitter uses the fieldbus to talk directly to a smart valve positioner. In others, a local microcontroller will handle the job of scanning I/O devices and then process inputs and outputs. Notice that the central CPU's role is to poll data from the microcontroller, compute control loops, monitor the system, and manage the network, not provide control *per se*. Odds are, there is no central CPU since informational functions can be performed from a workstation on-site or by a remote PC through the corporate intranet.

At this advanced phase of interconnection, many points of I/O are available to many nodes. The network becomes the controller. This approach is revolutionizing control networking.

For the control engineer, it is now possible to simplify the design, implementation, and maintenance of complex automation and control systems. With the many-to-many strategy, the engineer can reduce complex networks into manageable segments, distribute the control function, and then network different segments.

Managers using plant-floor workstations or HMI can pull information from a combination of devices to get not only a picture of I/O, but also a snapshot of key performance indicators to improve the operation. For executives, the same data available to the manufacturing execution system is accessible over the intranet or Internet to update the ERP cockpit for timelier decision making.

Finally, the many-to-many phase is leading to the many-to-any model. At this stage, the architecture becomes organic. Devices with their own Ethernet IP address can be moved and added as necessary, adapting to the changing needs of the operation.

An Open and Closed Case for Fieldbus

As seen in Figure 2–8, each data flow model fits with a corresponding network architecture. Each architecture requires increasingly higher degrees of interconnection, intelligence, and openness.

Fieldbus was born and matured in the era of all-to-one interconnection. Now in the era of many-to-any interconnections, it is beginning to show its age. That is because this level of networking—where innumerable devices and applications must communicate with each other—demands an open protocol for an open architecture.

The word *architecture* is often used when discussing networks, frequently in the context of **network architecture** referring to how nodes are wired together. But **architecture** can also refer to the overall structure of logical interrelationships between physical and/or conceptual components.

Openness—the ability of components made by different vendors to interoperate within a system—makes it possible for the network to tap the full potential of digital communication to improve data accessibility.

Openness is an issue in a plantwide network, because the network must connect with a number of high-level systems—SCADA, MES, MRP, data servers, supervisory PCs, firewalls—and with control devices from various vendors. It must also adapt to the needs among cell and plant supervisors, business system users, and network administrators.

Integration is also an issue, because management is pressing manufacturers to import data from control networks and plantwide networks into enterprise-level systems (Figure 2–9). After spending millions of dollars on IT initiatives, many companies are still paying people to type factory data into ERP systems by hand.

New business models require better data and systems integration to improve information flow as a way to boost competitiveness. In this scenario, the factory network is subsumed under a multi-level enterprise architecture.

It is fair to say that in the past, fieldbus was not very open. The capabilities of a given fieldbus were tied to a particular manufacturer's offering. You bought the manufacturer, and you bought their architecture.

For obvious commercial reasons, control system vendors did not want to connect devices to their bus. As openness became an issue in the 1990s, vendors responded by either publishing their standards or setting up nonprofit associations to let users decide how to implement their newly open fieldbus.

While addressing the issue of device compatibility and flexibility, this move fails to address the problems created by architectural complexity. As shown in Figure 2–10, the traditional hierarchical, multitier architecture can involve multiple gateways and middleware to handle routing and data conversion for different protocols used at each layer. Consequently, additions and changes to the network involve laborious reconfiguration and programming. Gateways and middleware software are not only expensive; they bog down performance. The entire message must be parsed and converted into Ethernet before transmission on the plant or enterprise network. Along the way, the same data may be stored in different formats in

Level 0 (Sensor level)	Connects discrete devices with limited intelligence, such as limit switches, photoeyes, or solenoid valves to controllers, PLCs, or PCs. It can also handle diagnostic messages originating from intelligent instruments, sensors, and I/O modules. The size of data is in bits, and the speed is in microseconds. Uses 4–20 mA signals, thermocouple voltages, and other "analog" signals. "Smart" field transmitters can also be used to communicate directly to compatible devices at any network level.
Level 1 (Device level)	Connects more sophisticated devices, such as controllers, that oversee I/O racks and I/O devices for variable-speed drives and HMI systems. Also connects smart devices that operate together in a distributed, time-critical network. Therefore, real-time response and "deterministic" data transmission are critical attributes at this level. Since device-level control handles data on the actual condition of the operation, not just a point of control, it is the data "feedstock" for all higher levels.
Level 2 (Fieldbus/Control level)	Links PLCs, DCS, and PCs with SCADA and HMI in a peer-to-peer network. The controlbus makes it possible to integrate data and exchange information so production units and manufacturing cells keep coordinated and synchronized. It also connects MES to enterprise management systems (MIS).
Level 3 (Plant level)	Connects the plant using a generic, plantwide data highway. Ethernet is commonly employed to connect PCs and workstations using multitasking operating systems—such as Windows NT, and Unix—with sockets for TCP/IP and other Ethernet network services.
Level 4 (Enterprise level)	Connects the factory network with the office LAN. These networks have fundamentally different requirements. In many cases, a secure "gateway" is necessary between the two networks (see Chapter 13 about firewalls or proxy servers). But security techniques must still allow information to pass from the enterprise to plant-level systems. Similarly, production data to determine available-to-promise status, inventory, and other resources must be made available to "front office" systems.
Level 5 (Internet level)	Provides information services to support the overall structure of the business and the data flow necessary for decision support systems, ERP, SCM, CRM, and e-Business.

FIGURE 2–9   Communication functions in a multitier network

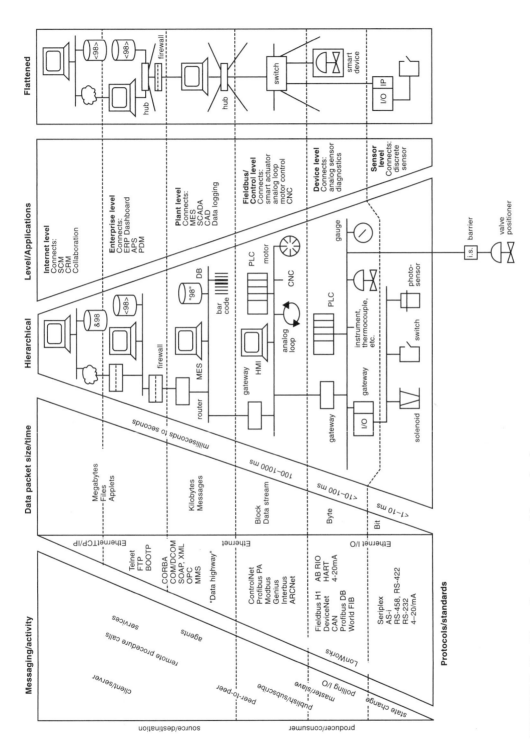

FIGURE 2–10 Tiers of control networking technology

different databases at each level—or never be converted and imported into higher-level applications at all unless it is entered manually. In other words, non-Ethernet protocols do not keep pace with evolving applications and technologies—a real speed bump on the information highway.

Fieldbus Fails Architecture 101

It would be nice if every factory network protocol and architecture could be seamlessly integrated. But in fact, decisions about control-system and factory-network designs are usually based on pragmatic issues like: What do we have to do to get the job done? How much will it cost? How long will production be down? How much of the existing system can we reuse? The result is haphazard network growth, a patchwork of proprietary and open protocols, and devices from multiple vendors—all of which make life interesting for the control engineer and control systems integrators.

HYBRID TOPOLOGIES

When a facility with a control-driven architecture must marry up with an information-driven architecture, the topology can become complex. The result is a hybrid topology that connects a closed, control-oriented bus topology with an open, information-oriented star topology. Star topologies are more complex to understand, but they are the topology of choice in corporate networks.

A star topology uses a central hub with cables radiating out to connect the nodes. Since each node can be connected to the hub by simply plugging in a cable, nodes can be easily added without interrupting the network. While failure of one connection does not affect other devices on the hub, failure of the central hub cuts communication with all nodes connected to it.

In most industrial networks, a hybrid topology can be used to create networks that accommodate hundreds, even thousands, of devices and systems. Unfortunately, fieldbus doesn't easily scale up to this size. And with maximum speed of 1 to 2 Mb/s, it struggles to meet the bandwidth required by real-time applications.

Fieldbus Fails as an Open Platform

Sensor, device, and control networks may employ any number of over 200 industrial protocols, a few of which are shown in Figure 2–10. Some are proprietary, some are open. But none are open to the extent that they will automatically deliver data to applications from Microsoft, Oracle, SAP, or Baan, i2, IBM, or Manugistics.

For example, when a high-level user requests device-level data—say, a reading of 98 on a water tank or 98 from a counter on a water-bottling production line—the reply must eventually employ the protocol used on the business network, namely Ethernet, and the data must be tagged in a format the application understands.

Although both factory and office networks employ digital communications, there can be a huge difference in their protocols.

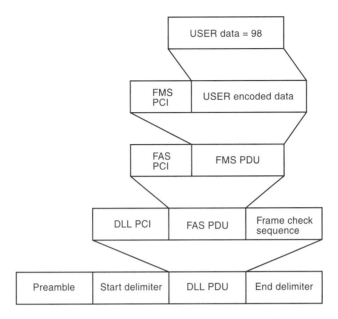

FIGURE 2–11   Nested fields in Foundation Fieldbus H1 to transmit 98

For example, Foundation Fieldbus is a popular process-industry protocol that allows scheduled (deterministic) and unscheduled access to the bus. (The term **Foundation** refers to the worldwide consortium of manufacturers and industry groups responsible for its development.)

As seen in Figure 2–11, the functions of fieldbus messaging are complex. As a contrast to Ethernet and TCP/IP, let's dissect how Foundation Fieldbus protocols handle the number 98. First, we will briefly examine the relevant layers.

Unlike TCP/IP, fieldbus packages network and transport functions in a unique fieldbus access sublayer (FAS). Various services and logical communication channels enabled by FAS allow different devices to communicate in a distributed communication system. For example, a **publisher/subscriber** service is used by a transmitter to send process variable (PV) data or a primary output (PO) to the controller and operator console. Once the communication link is established, the FAS protocol data unit (PDU) handles the fieldbus message specification (FMS) services that enable the actual data exchange. In general, FMS is responsible for the communication services, message formats, and protocol behavior needed to build messages for the user application. The FMS service of interest here is the variable access service (VAS) that allows the user application to access the variable associated with an object description.

Data communicated over fieldbus must use tags that function like HTML. These tags describe the data so they are properly formatted on the operator's console or in trend reports.

But fieldbus tagging differs from HTML in that formats are specific to the device. Fieldbus H1 employs device description language (DDL—a technique also used by HART and PROFIBUS PA fieldbuses). DDL is used to compose an **object description**—the format of the data communicated over the fieldbus. For convenience, object descriptions are collected together in a structure called an **object dictionary (OD).** The OD makes it possible to identify the appropriate object description by simply using an index number.

Index numbers below 256 define standard data types, such as an integer, float, or a bit string. In the case of 98, the float value is used to indicate a process variable.

The exact formatting of FMS messages follows a syntax description language called Abstract Syntax Notation 1 (ANS.1)—a standard developed by the International Telegraph and Telephone Consultative Committee (CCITT). For example, fieldbus H1 tagging to display a temperature value of 98 C looks like this:

```
VARIABLE Temperature
{LABEL "Temperature"
      TYPE FLOAT
      {DISPLAY_FORMAT "##.#";
      MAX_VALUE 99.9;
      MIN_VALUE 0.0:}
      CONSTANT_UNIT "° F";
}
```

Fieldbus H1 is a good example of complexity. The array of FAS, FMS, DLL, and OD components does not compare favorably with the friendly familiarity of TCP, IP, HTTP, and XML.

To aggravate the issue, portions of the fieldbus H1 protocol are assigned by the Foundation, and the entire fieldbus communication stack used by a device must be Foundation approved. Finally, the data itself—the number 98 is landlocked within tags that make it difficult to import into a database. To move the data to a higher-level database requires middleware or gateway hardware to strip the numerical value out of the FMS format, create a socket, and then send the data via Ethernet to the database application.

But these incompatabilities are not just a Foundation Fieldbus issue. We could just as easily picked on the DeviceNet and ControlNet suite, or on Profibus, Modbus, or Interbus protocols.

In any case, fieldbus incompatibility, data tagging differences, and hybrid architectures—all the result of what the control industry calls the "fieldbus wars"—complicate data accessibility.

Our goal of this chapter was to get the number 98 out of a device-level node to a data server at an information level, so it can be easily pulled into a Web page or onto an ERP dashboard. Due to the complexities of the fieldbus protocol, we failed.

The farthest we can go is to get 98 into a Fieldbus-compatible **data historian (DAQ)** at the control level. Gateway hardware, data conversion software, plus integrating applications—such as Datasweep—are necessary to finish the job. A number of problems would be solved if we could load the fieldbus telegram into an Ethernet packet's payload. Then the data could be delivered to compatible user applications, to Web, also to suppliers, customers, remote locations—to anyone who can legitimately profit by that information.

As will be shown in Part Three, a number of fieldbus organizations have devised ways to harness Ethernet as a universal vehicle for data delivery—taking a big step toward making data access faster, cheaper, and better.

SUMMARY

- Data flow is the fundamental reason for the existence of a control network.
- Control systems are employed to adjust, monitor, and record the condition or status of machinery, products, and processes.
- A closed-control loop uses feedback.
- Simplicity, economy, safety, flexibility, capacity, signal accuracy, and system reliability are the characteristics of a successful industrial network.
- Binary signals are a natural fit for discrete control, but create complications for analog control.
- The term *fieldbus* describes a large family of bidirectional, digital communication protocols.
- A full fieldbus protocol can handle byte-size data for complex transmitters and valves, plus diagnostics and control information.
- *Openness* means devices from different vendors can interoperate within the same system.
- Fieldbus fails as an open architecture.
- The demand for a universal data delivery has been an incentive for fieldbus organizations to combine Ethernet with their legacy, proprietary protocols.
- Users benefit as iE networks make data access faster, cheaper, and better.

? REVIEW QUESTIONS

1. What are the four components in control-system communication?
2. What is the difference between an analog and digital signal?
3. List three benefits of digital communication for control networking.
4. Explain the relationship between sensor- and device-level bandwidth requirements and discrete and process applications.
5. Why is real-time performance a major issue in an industrial network?
6. What are suitable network applications for HART and RS-232?
7. Where is a distributed control system located?
8. Why is it logical that the Fieldbus Foundation chose Ethernet to be the platform for their high-speed networking protocol—Fieldbus HSE?
9. What are the four different phases of industrial-network-architecture evolution?
10. What are the benefits of deploying an iE network in a *greenfield* installation?

The Business Value of Industrial Ethernet: To iE or Not to iE, That Is the Question

3

OBJECTIVES

After reading this chapter, you will be able to:

- Recognize the inter-organizational factors (market pressures and technology effects) that are driving industrial Ethernet (iE) to become the prevailing standard for plant networks.
- Identify the intra-organizational factors (business applications and manufacturing systems) that make an iE network a valuable part of a business's information infrastructure.
- Identify the intra-plant factors (such as plant optimization and control-network simplification) that determine the value of iE implementation in particular cases.

A good business depends on good communication. But as in personal relationships, people have doubts about whether "opening up"and "improving communication" are really worth the risk.

Therefore before considering the technical *whats* and *hows* of iE networking, it is worthwhile to examine the bottom-line *whys* of this technology's basic value. This perspective will be helpful when justifying the scope and shape of your particular iE implementation.

INTER-ORGANIZATIONAL FACTORS

Ethernet technology can carry information to points outside the plant or processing facility. As collaboration is increasingly important, we begin by looking at the benefits of networking from outside the organization.

Revolution of Rising Expectations

Unlike office Ethernet networks that replaced outmoded typewriters and PCs in a gradual evolution, deploying an iE network usually takes a revolution. The revolution may start with a simple phone call:

> We've got a customer buying three truckloads of widgets for $10,000. Can we deliver in 48 hours? Can we meet the price point? And can we give him an answer *now?*

These days, companies face customers who expect instant gratification in business information. Such expectations are largely due to the Internet, e-business, and related networking technology, which have dramatically accelerated the information flow among competitors, customers, suppliers, and allies. Instead of slow, scarce, and low-quality data, a network delivers data at a speed and depth that removes the informational friction between each group.

COMPETITORS

The Internet and computer-aided design and manufacturing technology now give competitors the ability to research, reverse engineer, redesign, and introduce cloned or improved products very quickly. To keep its competitive advantage, a company may need to shift to a lean or agile production environment, which requires new, real-time interconnections between business and manufacturing execution systems to provide relevant information throughout the supply chain.

CUSTOMERS

The ability to source goods and services over the Internet has raised customer expectations about product delivery, cost, and quality. This puts manufacturers in a double bind: On one hand they must engineer cost out of products to maintain margins; on the other, they must add value by improving product customization, service, and support. To meet both demands, a company may need a new information infrastructure that supports a **customer-relationship management (CRM)** solution to cut the cost of customer support while boosting customer loyalty.

SUPPLIERS

The burden to reduce material costs and lead times has shifted to the supplier, who often must manage inventory, implement **just-in-time (JIT)** delivery, add value to material design, and respond to unpredictable demand cycles. In return, a company must create interconnections that allow qualified suppliers to access sensitive manufacturing and customer data.

ALLIES

The need to become more agile forces companies to reduce organizational layers and to reassign personnel. To get services and technologies that were previously available in-house, a company must create interconnections that enable outside solution providers to function as part of the team. Internet or intranet connections may be used to deliver video training to redeployed personnel and machine instructions to inexperienced equipment operators.

Revenge of the Circular Network Effect

Like a high stakes poker game, Internet- and networking-related technology keep raising expectations—but companies keep hedging their bets. That is because businesses have been caught in a vicious and expensive circle: For nearly two decades, the computer revolution has liberated more and more information. Then the information revolution dictates that companies buy more and more computers. This cycle is called the **circular network effect.**

Figure 3–1 illustrates the exponential increase in communication transactions that results whenever businesses add interconnections.

To keep pace with this information explosion, complex networking technologies have coevolved to handle the data required by increasingly powerful business applications (Figure 3–2).

The advent of the personal computer in the mid–1980s opened a Pandora's box. Local area networks (LANs) made it possible for accounting and forecasting information to be shared within the financial department. Similarly, **computer-aided-design (CAD)** drawings and specification documents could be shared within the engineering department.

Then the advent of Novell networking in the early 1990s made it possible to share information interdepartmentally among finance, accounting, and marketing. Office LANs were then tied together to form wide area networks (WAN). By the late 1990s, the prevalence of PCs, Ethernet networking, and Internet and intranet access could leap geographic boundaries to network all departments and business functions—human resources, finance, operations, marketing, and sales—into the information enterprise.

But throughout this period, the connections between white-collar enterprise and blue-collar plant networks were slow in coming. Why? In general, change from the outside is not welcomed on the factory floor because it threatens to upset the way the plant has been run in the past.

Beginning in the mid 1980s, corporate facilities had the luxury of evolving from serial peer-to-peer, to Novell, to token-ring, to Ethernet networks. During that same time, manufacturing facilities built their own plant automation and control networks to meet their distinctive needs. Many existing plant floor controls and systems were installed during that period. One observer notes that many plants today are no more advanced operationally than they were a decade ago. For some, the biggest visible change is that operators now sit at desktop workstations rather than walking along a control panel board.

Plant managers may feel that a plant has plenty to lose from making networking "improvements." Retrofitting a different plant automation and control system can jeopardize production. Without a compelling reason, it just is not worth the risk. Change is only justifiable when the benefits outweigh the risks.

Exchanging four pieces of information

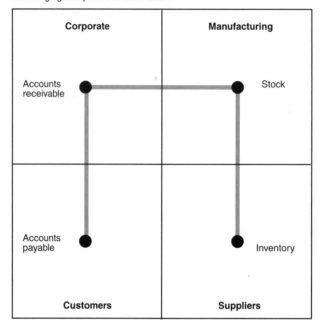

Exchanging twelve pieces of information

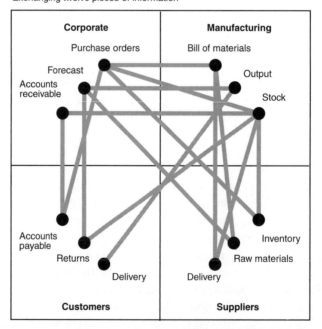

FIGURE 3–1   Network demands grow exponentially with increased demand for information

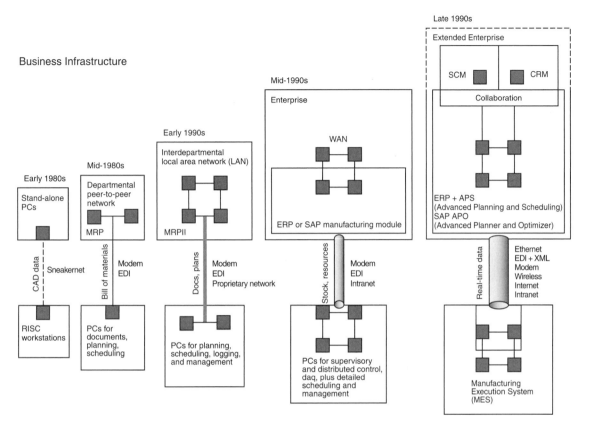

Business Infrastructure

Factory Infrastructure

FIGURE 3–2   Coevolution of systems, networks, and applications

So You Want to Start a Revolution?

By the late 1990s, the changes unleashed by the digital revolution created a series of cascading advances in computing technology that affected business strategies, manufacturing systems, product strategies, and procurement practices. These changes in turn have necessitated changes in the factory floor's information and network infrastructure.

Software developers triggered the avalanche by improving applications used to manage manufacturing. For example, for many discrete manufacturers, computerized **material requirements planning (MRP)** software is used to handle the ordering and the bill of materials for products with hundreds of parts.

Early in this decade, the interface between the corporate network and the factory network was relatively thin because the need to input factory information into office applications was relatively low (Figure 3–3). At this integration stage, operational command-and-control information required by computer-aided process planning and MRP was (and often is) input by hand.

	Level 1: Command-and-control integration	Level 2: Transactional integration	Level 3: Collaborational integration
Corporate IT	• MRP or MRPII with partial (financial) ERP • Database/Servers • Internet • E-mail • Passive Web site • DSL/Cable/T1 fractional • Peer-to-peer LAN with hubs	• MRP II/ERP • E-commerce • Trading exchange • Multiple databases • Internet • Intranet • E-mail • T1–T3+ • Interactive Web • Switched network within building	• ERP/SCM/CRM • Trading exchanges • Common database • Internet • Intranet • E-mail • OC-3+ • DHTML Web • WAN • VPN/VLAN
Plant IT	• Manufacturing information database • Legacy control networks • CAD/CAM	• MRPII • Supervisory control and remote control • Legacy plant automation network • CAD/CAM/CAE • Bar code	• MES • SCM/CRM gateway • Wireless • Remote monitoring and diagnostics • CIM • Open network • PDM • Bar code
Application	Control network interfaces with a plant information system—MPC, MES, APS. Report data available to MRP or ERP thru manual input/e-mail.	Plant's internal supervisory systems can send recipes, set points, and targets to controllers for automatic implementation. Report data available to MRP II thru middleware or an ERP thru APS bolt on.	MES links with external systems (SCM/CRM) to provide *real-time* information on inventory, order, WIP, yields, and KPI indicators. External systems (via Internet or intranet) can send recipes, set points, and targets to controllers for automatic implementation.

FIGURE 3–3   Three phases of business/manufacturing network integration

- Sales, customer, and order demand-related information
- MRP: Component planning and part data
- Inventory status data: Bar code
- MRPII: Process information
- Internal control and security access tables for client/server
- Cost collection: Standard-actual-activity costs
- Performance measurement extracts
- Customer information
- Customer satisfaction information
- Stockholder and treasurer information
- Vendor and supply-chain related data
- Employee HR data

FIGURE 3–4   Information contained within an ERP system

But software developers went on to develop PC-based **manufacturing resource planning II (MRPII) applications.** An MRPII system allows purchasing, marketing, engineering, finance, and accounting to use outputs from MRP to coordinate work and achieve cost savings.

As the capabilities of MRP and MRPII software grew, these applications took advantage of networking and database technology to draw upon inventory, resource, and operation information from the factory floor.

By the late 1990s, MRPII developers introduced **enterprise resource planning (ERP)** applications. ERP scratched the lust for instant gratification by promising to make all key business information available in a common database (Figure 3–4). Eager corporate customers created an ERP boom in 1999 and 2000, which went bust in 2001.

ERP proved to be problematical. Not only did it cost a fortune to integrate data from many different sources, ERP upset the way the plant had been run in the past and provided few tangible benefits to the factory. Plus, ERP applications, which began as MRPII *forecasting* tools for high-volume discrete manufacturing, were deficient in handling *immediate* information that was relevant to process and build-to-order manufacturing. Specifically, ERP failed to handle actual operational data required by customer-oriented **pull** manufacturing systems, **supply chain managements (SCM)** applications, and e-business business-to-business (B2B) transactions requiring up-to-the-minute information.

Most small to midsize companies retreated from expensive ERP and e-business projects and looked for initiatives with a clear **return on investment (ROI).** For example, some industries have successfully established private or public trading exchanges to procure lower-cost materials. For these companies, the value of an iE network is more for communication and collaboration rather than e-commerce.

Many larger companies, however, have devoted additional resources to rescue ERP. They succeeded either by successfully bolting on **advanced planning and scheduling (APS)**

software to remedy ERP's inability to handle trend data, by adding SAP's **advanced planner and optimizer (APO)** modules, or simply by using a more capable ERP provider. Today, these companies are positioned to take full advantage of not only trading exchanges, but also developments in SCM, CRM, and Internet-enabled collaboration.

At great expense, these companies have integrated information applications, manufacturing systems, production strategies, and network infrastructures to bridge the gap between the plant network and the corporate enterprise. This extended enterprise requires a very fat network pipe to interconnect manufacturing, corporate, supplier, and customer enterprises to the information stack required by SAP, ERP, SCM, and CRM applications (see Figure 3–2).

Information systems share this data with internal applications, as well as through external Web front-end applications and **business-to-business (B2B)** exchanges. Many of these exchanges and transactions must occur in "real time," meaning in this case, a time period that satisfies the user. For example, if a production line suddenly breaks down, all members can be alerted in time to make appropriate adjustments.

B2B communication links depend on connecting Web-facing storefronts, back-office databases, ERP, and MES systems so transactions initiated by a browser or wireless handheld client can cascade in real-time through the entire chain of applications. Networking technology is essential for carrying operational information in real time from different locations in the chain to users who need to know what is happening right now.

According to the Gartner Group, most information-integration projects are driven by the need to:

- Increase the information available to senior management.
- Support e-business initiatives.
- Adapt to the increasing rate of change.

The goal is to create a network architecture that supports all kinds of interactions between business partners and customers with functional processes and the information infrastructure inside the enterprise.

INTRA-ORGANIZATIONAL FACTORS

At this point, it is important to dispell the notion that the type of business that most requires iE networking will be a *wired* e-business catering to e-commerce customers. Carrying information between an enterprise and its factory network is just as challenging as collaborating with the outside world.

Network Fit with Manufacturing Systems and Production Strategies

Traditional manufacturers still have reasons to adopt iE networking, because it is an infrastructure that supports such efficient manufacturing concepts as lean manufacturing, **build to** order **(BTO)** or **available to promise (ATP)** production strategies, **pull scheduling, flow manufacturing, cellular organization,** and supply-chain management. All these techniques predate the rise and fall of *dotcom* initiatives. They just happen to be valid approaches that benefit from the Internet and the network capabilities of iE, because the success of each of these concepts is predicated on greater information flow.

Eliminating Information Bottlenecks

Deploying an iE networking is a move to eliminate information bottlenecks between the plant floor and corporate information systems. At the same time, iE facilitates the exchange of information between ERP business and MES manufacturing systems to respond instantly to CRM and SCM demands (Figure 3–5). Thus, a key selling point of an iE network is that it is mission critical as part of an information infrastructure that can respond to the pressures and potentials of today's ever more dynamic, network-driven marketplace. In contrast, consider the type of manufacturer who finds it difficult to justify iE.

Many build-to-stock mass producers or bulk commodity suppliers fall into this category (Figure 3–6). Because information barriers are part of their business model, they think they can get by with proprietary plant control networks and fieldbuses—which is logical, because such businesses may not have the margins and capital to afford costly ERP or e-business technologies.

In these cases, the factory is treated like a black box—production goals go in and product comes out. Information inputs include bills of materials, parts, and scheduling. Outputs include finished goods and wastage. For example, any miscommunication between the factory and management can be handled by warehousing excess inventory, drawing stock from other distributors, or simply backordering customers.

Now consider the type of manufacturer who is an eager candidate for iE. This may be a discrete manufacturer with a build-to-order production strategy or with an adaptive manufacturing system to achieve mass customization. In the process industry, a good case would be the manufacturer who handles custom batch orders and custom recipes. Both types have a lot of incentives to eliminate information barriers.

An ERP system that employs APS, which also functions as a core part of an SCM application, will demand real-time information flow over the network. The logic of APS needs real-time data to optimize the actual execution of manufacturing (MES) operations—which is critical to build-to-order production strategies and to minimization of **work in progress (WIP).**

For both the build-to-order discrete manufacturer and the custom-batch processor, iE networking provides the bandwidth that can handle multiple, simultaneous users for extensive reporting, monitoring, data uploads, and instruction downloads.

For these reasons, it is obvious that iE networking should fit into the factory IT plans of a BTO computer maker or pharmaceutical manufacturer. Other obvious fits: a lights-out **computer-integrated-manufacturing (CIM)** facility that practically operates without direct human intervention, and a virtual manufacturing venture that harnesses small companies together for a short term to function as a larger manufacturer to produce a single product.

But there are many other manufacturing systems and production strategies where an iE network proves valuable.

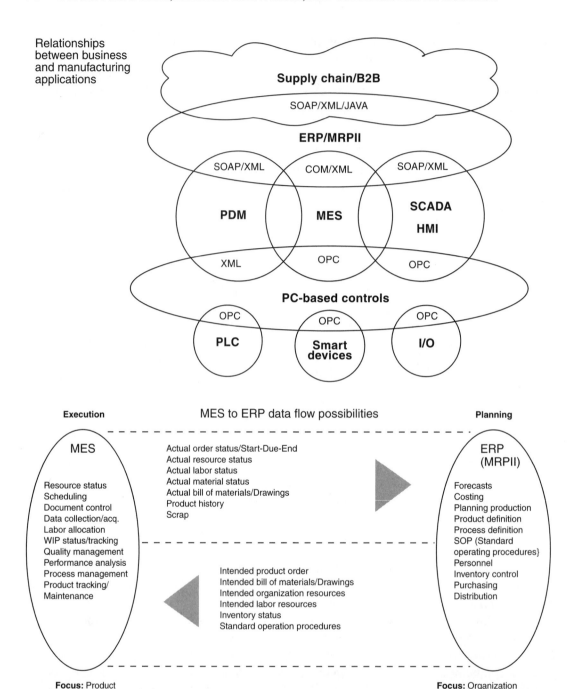

FIGURE 3–5   Information flow between MES and ERP systems

Business characteristic	Feature that makes it unlikely to value iE	Feature that makes it likely to value iE
Plant-to-enterprise network interface	Thin pipe	Fat pipe
Production strategy Discrete: Process:	Make-to-stock Bulk commodity	Make-to-order Custom batch
Lot size Discrete: Process:	EOQ (Economic order quantity) Bulk	One Custom recipe
Planning	Forecast driven	Demand driven
Demand	Push	Pull
Plant layout	Job process	Product flow: Cellular, FMS
Inventory	JIC (Just-in-case)	JIT/VMI
Manufacturing system	Mass production	Mass customization (Lean>Agile)
WIP	High	Low
E-apps	N/A	SCM/CRM/E-CRM
Planning apps	MRP or MME	ERP + APS
Factory network	Proprietary fieldbuses and protocols	Open Ethernet and TCP/IP

FIGURE 3–6   Business models least likely and most likely to value iE

Networking in Flow and Cellular Manufacturing and in Processes

We already considered the slam dunk case for iE networking: a BTO or ATP production strategy employing fixed-automation or flexible-automation systems in continuous or line manufacturing (Figure 3–7). Such a production environment depends on real-time operational data, as well as a lot of data on parts and distribution of finished goods.

FLOW MANUFACTURING

There are other cases where real-time data flow—in the specific sense of keeping up with millisecond scan times—isn't an issue. For example with **flow manufacturing** and JIT

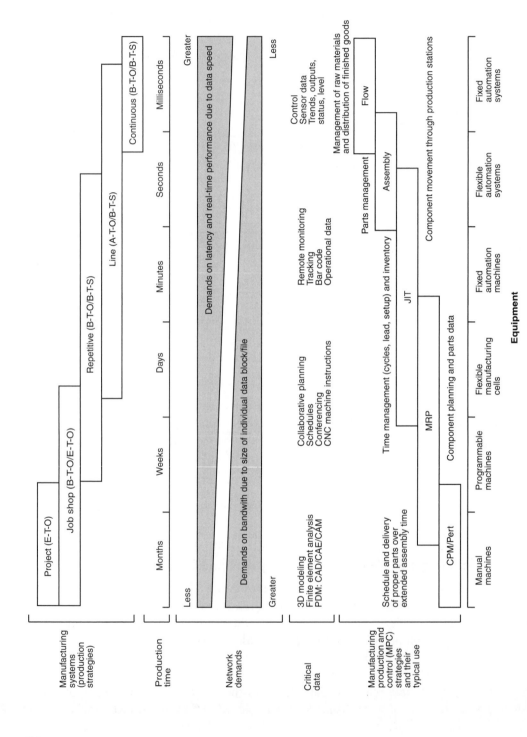

FIGURE 3–7 Production environment variables that place demands on an iE network

inventory control, a wide range of information must flow at a pace that keeps up with the drumbeat of manufacturing steps required to synchronize cycle times and material flow. Similarly with job shop and project manufacturing methods, the network is used more for carrying information for collaboration with members of the production team. In these cases, extensive bandwidth may be needed for **product data management (PDM)** applications— **computer-aided design, engineering,** and **manufacturing** (CAD/CAE/CAM)—computer modeling, and finite element analysis used by collaborative engineering and design teams.

CELLULAR MANUFACTURING

A flexible **manufacturing cell** is often used in conjunction with flow manufacturing. In general, flow manufacturing concerns velocity of operations and the material stream to minimize interruptions; cellular manufacturing concerns flexibility of operations and the organization of the plant floor. Together, they are part of implementing a leaner manufacturing system that cuts the cost of inventory, setup time, and lead time.

Consequently, a cellular approach puts all the resources in one spot to perform all the operations needed to produce a family of parts. Resources may include computerized-numerical-controlled (CNC) machines, programmable logic controllers (PLCs), bar code readers, automated material handling (automated guided vehicles [AGV] and robots), and a storage system.

While the MES controls the entire factory, the cell controller handles the activities of the cell—which is extremely information intensive. Communication involves program control for managing production, scheduling, resource availability of materials and devices, CNC programming, monitoring of cell and device status, error monitoring, changeover engineering specifications, and production tracking. Consequently, over 50% of network utilization by the cell controller involves large, megabyte file transfers (Figure 3–8). A fieldbus system does not have the bandwidth to handle the information generated by entire manufacturing cells, but an iE network does.

Standardizing on a common iE protocol stack means fewer gateways and data-integration hassles. It facilitates sharing information across a wide range of applications employed on the plant floor, R&D labs, business systems, and other parts of the enterprise (Figure 3–9).

PROCESS OPERATIONS

In the process industry, off-the-shelf software can marry SCADA systems to business systems. Executives can then see trend information, not just throughput data, from a browser window. This capability also saves operators from walking across the entire facility to monitor a process.

A new business has emerged to take advantage of remote process monitoring, called **remote diagnostics.** With remote diagnostics, industrial systems and components can be serviced and maintained from off premises. Operational data, including alarms and faults, are collected in a secure database. The data are then analyzed over a secure wireless or Internet connection for abnormal or less than optimal conditions. In some cases, corrective and repair

Communication level	Tools/ applications	Users	Best response time	Content		Media	Network
				Information	Data		
Extended enterprise	Browsers Portals Exchanges Commerce Remote diagnostics	1,000s	minutes	95–100%	0–5%	fiber	Internet Intranet
Enterprise	PCs ERP APS SCM CRM	100s	less than a minute	85–95%	5–15%	UTP	WAN firewall
Plant network	PCs PDM (CAD) MES MRP	10s	seconds	70–85%	15–30%	UTP fiber	LAN broadcast domain
Cell/supervisory control	HMI SCADA PLC DAQ	under 10	milliseconds	15–70%	30–75%	UTP wireless fiber	switched network
Device	PLC I/O	individual	milliseconds microseconds	5–15%	85–95%	UTP fiber	single collision domain

FIGURE 3–8 Levels of network utilization

Application	Bandwidth	Real time	Prioritization
APS	+	++	++
Analysis (FEA, etc.)	++	– –	0
Bar code	0	+	+
Computer modeling	++	– –	0
Control	++	0	++
CNC	+	++	++
Document sharing	+	–	0
E-mail	– –	– –	0
ERP	varies	varies	++
Extranet	++	0	++
Internet	+	+	varies
Intranet	+	+	++
Operations	++ to +	++ to +	++
PDM (CAD/CAE/CAM)	++	–	0
SCADA/remote monitoring	+	+	+
Sensor	– –	++	++
Simulation	++	+	0
Video conferencing	++	++	++
Voice/video/data convergence	++	++	++

FIGURE 3–9   Network utilization by productivity application

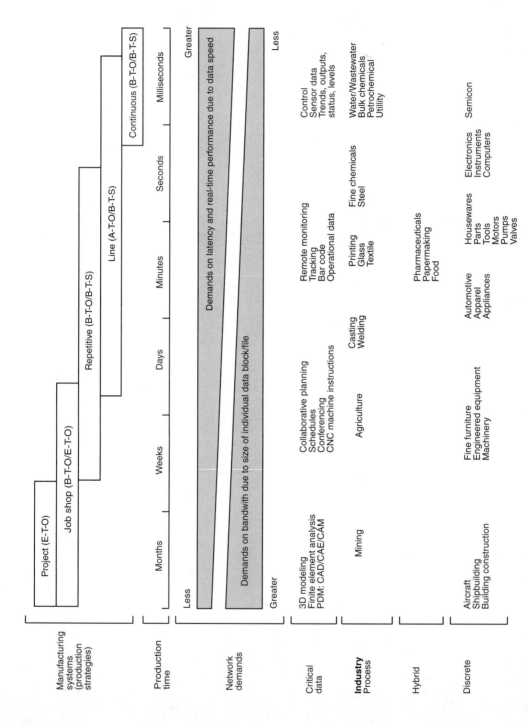

FIGURE 3-10 Network utilization by industry

programs can be directly uploaded from the equipment manufacturer into the machine control's firmware. In other cases, appropriate product documentation can be easily supplied to the customer or field service technician.

To help you evaluate the utilization of an iE network for a given manufacturer, the array of variables in Figure 3–7 can be viewed by industry, as seen in Figure 3–10. (While network load correlates directly with applications and users, by extension it can also be loosely correlated with industries that tend to use certain apps.)

For a wide range of industries, an iE network proves its value by handling bandwidth, prioritization, and real-time demands required by different, data-intensive applications. Specifically, iE has the versatility to support the gigabit/second or higher capacities needed for voice and video streams that may be required for equipment operator training, plus the performance and cost effectiveness to handle device-level demands.

INTRA-PLANT FACTORS

The production environment presents a wide array of variables—equipment configuration, management system, production strategy, production time—that place demands on the industrial network. But the bottom line is saving manpower and money.

Saving Time

For example, a machine operator downloading instructions to a CNC machining center or a control engineer uploading recipes to several PLCs is wasting time if the network is bogged down. A properly segmented iE network has ample capacity to handle I/O updates involving lots of bits and files with loads of bytes.

And compared to laborious configurations of fieldbus devices, an iE network can be commissioned quickly. Depending on the application, network configuration can be as easy as using **dynamic host configuration protocol (DHCP)** to assign IP addresses automatically.

Saving Money

In addition to saving time, an iE network can simply save money. When compared to existing plant automation and control networks, it is more economical in several critical areas.

COST OF COMPONENTS

Because Ethernet components are competitively produced by a number of manufacturers instead of a proprietary source, they are relatively inexpensive. Compare a $1500 proprietary I/O adapter with a $10 Ethernet adapter. (Caveat: If a number of switched hubs are necessary to create many high-speed, full-duplex, noncollision domains, cost of switched hubs and routers can become an issue. Furthermore, additional cabling, switches, and firewalls may be needed for redundancy and security.)

COST OF TRAINING

Because in-house personnel are probably already acquainted with personal computers that use Ethernet adapters and compatible protocols, such as TCP/IP, Ethernet has a lower learning curve. When using Ethernet to replace multiple proprietary network protocols, it simplifies training and support.

COST OF WIRING

Complex point-to-point wiring can be dramatically reduced with a distributed control system that can replace miles of wires with a few Ethernet cables and hubs or switches. If it becomes necessary to relocate equipment or sensors, running wire to reconnect the device to the Ethernet backbone is inexpensive and as simple as running one cable from the device to a nearby hub.

COST OF ARCHITECTURE

A PC-based control system can create one network layer from sensor to supervisor by eliminating tiers of I/O modules, PLCs, workstations, and supervisory computers. When deployed in conjunction with PC-based control, the PC compatibility and economy of Ethernet adapters makes iE the compatible, cost effective network specification.

Case-by-Case Considerations

Continuing economic pressures in the manufacturing sector make iE a strong competitor to existing industrial network protocols—but is it the logical successor? Many industrial control system integrators are acknowledging that it is. A recent study done by Industrial Controls Consulting (Fond du Lac, WI) shows that U.S.–based system integrators selected Ethernet by three to one over the closest rival—the decade old industrial protocol named DeviceNet. In further recognition of Ethernet's ascendancy, the sponsoring organization for DeviceNet (ODVA) issued in March 2001 a 400–page "Ethernet/IP" standard, which allows DeviceNet's CIP protocol to be packaged inside an Ethernet TCP/IP frame.

Even if iE is the wave of the future, you need to consider what place it will hold in yours. In some cases, the decision to implement an iE network will simply be a business necessity dictated by an ERP or SCM solution that requires end-to-end, real-time data availability. In others, a full build out of iE's voice, video, data, wireless, and Internet capabilities will be needed to provide a full range of communications across the entire enterprise. In others, a decision will be reached merely to deploy iE tactically, for example, to reduce the complexity and cost of maintaining multiple legacy control networks. In still others, the network will be used for remote diagnostics to take advantage of the expertise of techs and OEMs off premises. The decision to move to iE as an open, high-bandwidth, industrial network may also be a smart, incremental step that strategically positions the plant for future business opportunities.

GREENFIELD SCENARIOS

As seen in Figure 3–11, full iE network implementation should be considered for a **greenfield** manufacturing plant, that is, a completely new facility, where it can be installed without the expense of a retrofit. It is the logical choice where a flat, PC-based control architecture is

Situation	Source of cost savings	Best time to deploy for best ROI
Greenfield Plant: Install end-to-end iE PC-based control	Save on flat hierarchy on wiring and avoid PLC costs	From the get-go
Greenfield Plant: Install iE at supervisory network level or for Distributed Control System (DCS)	Save on wiring and cost of PLCs; provides real-time redundancy	From the get-go
Brownfield Plant: Upgrade to iE at supervisory network	Expense offset by savings on wiring and maintenance	Fair ROI—advisable with ERP upgrade
Brownfield Plant: Retrofit to end-to-end iE PC-based control	Huge expense partially offset by savings on PLCs, wiring, and maintenance	Questionable ROI—advisable with plant floor layout reorganization or process re-engineering
Brownfield Plant: Add iE for remote SCADA	Expense offset by saving labor of remote site visitation; improved trend logging and response	Good ROI—deploy anytime when minimally disruptive

FIGURE 3–11   ROI relative to business situation

used for control in lean or agile manufacturing and where information intensive BTO and ATP production strategies are employed.

BROWNFIELD SCENARIOS

In certain circumstances, full iE network implementation may be appropriate for a brownfield plant retrofit. For example, a CIO may be called upon to rescue ERP by remedying the gap between it and the MES. A software fix may involve bolting an APS module onto the ERP application. An appropriate iE network upgrade may also be needed to carry the load of data generated by manufacturing cells.

Another reasonable retrofit opportunity is deploying iE as an added value feature when replacing outmoded proprietary control architecture with an open, PC-based control system.

For collaboration-intensive production environments, an investment in iE is also justified to handle remote monitoring over the Internet, bandwidth intensive modeling applications, and voice/data/video streams. And in a transactional environment for private or public trading exchanges, iE enables VLAN (virtual local area network) and VPN (virtual private network) connectivity for SCM and VMI (vendor managed inventory) applications.

During iE network decision-making it is important to include representatives at every level of the organization, from the machine operator who uses the control panel to the manager of e-business strategies. The buzzword is *team*. Plant control and corporate IT engineers need to ensure that the performance and security of their respective networks enterprise are not compromised. But just as importantly, factory and business team members need to look at the information infrastructure decisions in general and networking choices in particular to achieve complementary business and manufacturing objectives.

SUMMARY

- IE technology is in step with the extension of the enterprise and Internet for collaboration with customers, suppliers, and offices.
- Only iE has the bandwidth to carry the information and operational data between SCADAs, PLC, data historians, and ERP systems.
- IE is a scalable pipeline that can handle byte-sized data packets at the device level as well as mega- and gigabyte demands for 3D modeling, video, and block transfer from manufacturing cells.
- IE facilitates popular lean, agile, and adaptive production environments by providing plug-and-play connectivity.
- Moving to iE may be more affordable with legacy protocols that are forward compatible with Ethernet (EtherNet/IP, Modbus TCP/IP, etc.).
- Choose iE for low cost of components, deployment, and training.
- Consider iE from the start for greenfield installations.
- Brownfield deployment of iE is usually justifiable only incrementally.

? REVIEW QUESTIONS

1. How does the networked-enabled business paradigm benefit competitors, suppliers, customers, and allies?
2. Why must companies increasingly resort to deploying more computer- and networking-related technologies?
3. Give three reasons companies embark on application integration projects.
4. Why does ERP necessitate a network-intensive information infrastructure?
5. What type of plant layout is more likely to need the high bandwidth of iE?
6. What types of manufacturing are most likely to need the real-time performance of iE?
7. What business planning application is most likely to need prioritized data delivery?
8. What are the savings of deploying iE? What can add to the expense?
9. In retrofitting iE in a brownfield facility, what application is most cost effective?
10. Why is an iE network particularly appropriate for a lights-out CIM facility?

PART TWO

Ethernet:
The Foundation
Protocol

The Foundations of Ethernet 4

OBJECTIVES

After reading this chapter, you will be able to:

- Define the three parts of Ethernet's naming nomenclature.
- List the three main Ethernet data rates most appropriate to industrial networking.
- Describe the parts of an Ethernet frame.
- Define the access method used by Ethernet.
- Describe why CSMA/CD protocol makes Ethernet nondeterministic.
- List the seven layers of the OSI model.
- Fit Ethernet into the OSI model.

ETHERNET: THE NETWORK FOR ALL REASONS?

The story of Ethernet is the story of a network technology that has evolved to meet growing and changing demands. The main demand is more bandwidth—that is, high performance through greater speeds. Ethernet began as a 10-Mb/s network. Today, it runs one thousand times faster, pushing 10 Gb/s.

Yet speed alone does not account for Ethernet's success. In other ways it also evolved. For example, it was one of the first networks to migrate from coaxial or other expensive, hard-to-work-with cables to inexpensive unshielded twisted-pair cable. It evolved from a bus structure to a more reliable star-wired topology. It provided redundancy, such as double power supplies, for more reliable operation. To be sure, other networks also evolved, but none evolved so fast and flexibly.

Another consequence of Ethernet's success was high-volume application. High volume drove down the price of components. Just as personal computers became less expensive and more powerful every year, so Ethernet's evolution ensured continued improvements in performance at lower costs.

Origins

Ethernet was invented by Bob Metcalfe and a team of engineers at Xerox's famous think tank, the Palo Alto Research Center (PARC), in the 1970s. The original network ran at 2.94 Mb/s. Soon, Xerox joined with Intel and Digital Equipment Corporation to provide a standard for 10-Mb/s Ethernet. This standard is called DIX (after Digital, Intel, and Xerox). The last version of this standard was DIX 2.0. The DIX version of Ethernet ran over thick coaxial cable.

At about the same time, the Institute of Electrical and Electronic Engineers (IEEE) was working to establish standards for local area networks (LANs) and metropolitan area networks (MANs). The LAN/MAN standards were developed by the 802 task force, with several committees working on different standards. Ethernet was developed by the 802.3 committee. Other committees developed other standards. Here are some examples:

802.2:	Logical link layer
802.3:	Ethernet (Carrier-sense, multiple access with collision detection)
802.4:	Token bus
802.5:	Token ring
803.6:	Distributed queue dual bus (DQDB)
802.11:	Wireless Ethernet
802.12:	100VG-AnyLAN (Demand priority)
802.17:	Resilient packet ring

Today, only 802.3 (Ethernet) and 802.11 (wireless Ethernet) are vital, growing standards. (802.17 is not properly a network; it is a topology of structures to ensure reliable, robust communications by providing a mechanism to allow a wide area network to recover quickly from errors.) While 802.5 (token ring) enjoyed a run at competing with Ethernet and 802.12 (100VG-AnyLAN) was an interesting technology with potential as a contender, they have fallen behind. In office networking, Ethernet is the only game in town. Another network option, called asynchronous transfer mode (ATM) has found success in wide area and telecommunications applications, but has not succeeded in becoming popular in LANs.

The first 802.3 standard, adopted by the IEEE in 1983, was similar to DIX 2.0: 10 Mb/s over thick coaxial cable. Officially, Ethernet is distinguished by its access method: carrier-sense, multiple access with collision detection. Indeed, the official title of the standard is "Carrier-Sense, Multiple Access with Collision Detection and Physical Layer Specifications." The name *Ethernet* is not used in the standard. Since then, over twenty versions of Ethernet have been created by the committee. Indeed, an argument can be made that Ethernet properly applies only to the DIX standard and not to the IEEE standards. However good the argument is, everybody ignores it.

The different versions of Ethernet are sometimes distinguished by a letter and date appended to the standards number. For example, IEEE 802.3u-1995 covered 100 Mb/s over twisted-pair cable, while IEEE 802.3ac-1998 covers virtual LANs. Occasionally, the committee gathers all the revisions together and issues a comprehensive standard. The latest is 802.3-2000 (6th edition), running 1515 pages, 43 chapters (called *clauses* in the standard), and 36 appendices (termed *annexes*).

Ethernet is an international standard. The IEEE is part of the American National Standards Institute (ANSI). ANSI, in turn, represents U.S. interests in the International Standards Organization (ISO) and International Electrotechnical Commission (IEC). The IEC and ISO have adopted the 802.3 standard as ISO/IEC 8802-3.

What about Wireless Ethernet?

A wireless version of Ethernet has been developed by the 802.11 committee. The reason it was developed by another committee is that it uses a different access method. This method is called **carrier sense, multiple access with collision avoidance (CSMA/CA).** We will discuss wireless Ethernet in some detail in Chapter 8. Right now, we are concerned with 802.3 Ethernet.

A QUICK ETHERNET OVERVIEW

Figure 4–1 summarizes the different "flavors" of Ethernet. The main variations involve differences in transmission rates and the type of cabling that the system runs on. The versions most important to factory applications are shown in **boldface.**

A quick explanation of the terminology for naming the different versions is needed. Notice that each flavor was a three-part name in the form of prefix-base word-suffix. In 10BASE-T, for example, *10* is the prefix, *BASE* is the base word, and *T* is the suffix.

The prefix number refers to the speed of the signal:

1 Mb/s (yes, Ethernet even backslid to a slower version, also known as StarLan)
10 = 10 Mb/s
100 = 100 Mb/s
1000 = 1 Gb/s
10G = 10 Gb/s

The middle or base word refers to the type of signaling, baseband or broadband. Only one version of broadband Ethernet has been created, and it has not become popular. The suffix tells something about the cabling medium. For coaxial cable, it is the segment length. For others, it is typically the type of cable.

Flavor	Cable type	Segment length* (meters)	Year of IEEE ratification	Speed
1BASE-5	UTP	250	1987	1Mb/s
10BASE-5	Thick coax	500	1983	10 Mb/s
10BASE-2	Thin coax	185	1985	(Ethernet)
10BROAD-36	Coax (broadband)	3600	1985	
10BASE-T	Cat 3 UTP	100	1990	
10BASE-FP	Fiber	1000	1993	
10BASE-FB	Fiber	2000	1993	
10BASE-FL	Fiber	2000	1993	
100BASE-TX	Cat 5 UTP	100	1995	100 Mb/s
100BASE-T4	Cat 3 UTP (4 Pairs)	100	1995	(Fast Ethernet)
100BASE-T2	Cat 3 UTP (2 Pairs)	100	1997	
100BASE-FX	Fiber	2000	1998	
1000BASE-T	Cat 5 UTP	100	1999	1000 Mb/s
1000BASE-CX	Coax	25	1998	(Gigabit Ethernet)
1000BASE-SX	62.5/125 Fiber	260	1998	
(850 nm)	50/125 Fiber	525	1998	
1000BASE-LX	62.5/125 Fiber	550	1998	
(1300 nm)	50/125 Fiber	550	1998	
	SMF	3000	1998	
10GBASE-SR	MMF (850-nm serial)	65	2002	10 Gb/s LAN
10GBASE-LX4	MMF(1300-nm WDM)	300	2002	(10-Gigabit Ethernet)
10GBASE-LR	SMF (1310-nm serial)	10,000	2002	
10GBASE-ER	SMF (1550-nm serial)	40,000	2002	
10GBASE-SW	MMF (850-nm serial)	65	2002	10 Gb/s WAN
10GBASE-LW	SMF (1310-nm serial)	10,000	2002	(10-Gigabit Ethernet)
10GBASE-EW	SMF (1550-nm serial)	40,000	2002	

*Segment length is the backbone cable length for 10BASE-5 and -2. For others, it is the distance from hub to station (or other attached device). Gigabit Ethernet distances over fiber depend on the type of fiber and transmitter used. Specifications for 10-Gigabit Ethernet are preliminary and not yet approved by standard-setting groups.

FIGURE 4–1   Evolution of Ethernet standards

Notice one other thing about the evolution of Ethernet shown in Figure 4–1. Over time, it has speeded up. The time from the original 10-Mb/s Ethernet to 100-Mb/s Ethernet was twelve years. Only three years were required to move from 100 Mb/s to 1 Gb/s, and 10 Gb/s has followed quite rapidly.

THE ETHERNET FRAME

In Chapter 1, we examined the code contained in different frames inside the data packet. Here we look at the specific fields inside the Ethernet frame. There are eight fields within the standard Ethernet 802.3 frame, which contain information about the source and destination synchronization information so that the receiving device is ready to receive the data (Figure 4–2).

Field Functions

PREAMBLE
The preamble is 56 bits of alternating 1s and 0s. 10-Mb/s Ethernet uses the preamble to synchronize between stations. It also allows a receiving station to lose a few bits as it reaches synchronization without loosing valuable information.

START FRAME DELIMITER
The start frame delimiter is 8 bits. The first six continue the alternating 1s and 0s of the preamble. The last two bits are 1, 1, which signal the receiving station that the next field is coming. Fast Ethernet and Gigabit Ethernet, because they use complex encoding schemes for the information, do not require this synchronization process. While they send the preamble and start frame delimiter, they do so for purposes of compatibility and not because they use them.

DESTINATION ADDRESS
The destination address is the address of the receiving station, in 48 bits. All addressable Ethernet equipment has a MAC address, assigned at the factory. (You can change this address, but doing so is only for the intrepid expert or the foolhardy novice.) The transmitting station can use a **unicast** address, a **mulitcast** address, or a **broadcast** address. The unicast

Preamble 7 bytes	Start Frame Delimiter 1 byte	Destination MAC Address 6 bytes	Source MAC Address 6 bytes	Length/ Type 2 bytes	Data 0–1500 bytes	Pad 0–46 bytes	Frame Check Sequence 4 bytes

FIGURE 4–2   Eight fields in standard Ethernet frame

address is the address of a specific station. A multicast address is an address that more than one station has been enabled to receive. Finally, a broadcast address means that all attached stations should receive it.

SOURCE ADDRESS
The source address is the 48-bit address of the transmitting station.

FIELD LENGTH OR FIELD TYPE
This sixteen-bit field can convey two types of information. If the value of the field is less than 1518 (the maximum number of bytes in a frame), then the field is being used to give the number of bytes in the data field. If the number of data bytes is less than the minimum required, then additional pad bytes will be added. The field length allows the receiving station to distinguish between real data and padded bytes added. If the value is 1518 or greater, the field is being used to specify the type of protocol data being carried in the data field.

DATA
The data field carries the actual data. It must contain between 46 and 1500 bytes to be a valid field. If the actual data is less than 46 bytes, then the transmitting station must add additional padding to bring the field up to 46 bytes.

The standard also allows logical link control (LLC) data to ride in the data frame. The LLC can provide control information or it can be used to identify the type of protocol used by the data. Notice that both LLC information and data can reside in the same frame.

FRAME CHECK SEQUENCE
The frame check sequence ensures that the received frame is the same as the transmitted frame. As the transmitting station sends each byte, it performs a mathematical operation based on the contents of the field (except the preamble and start frame delimiter). It enters the result in this frame. The receiving station performs the same mathematical operation and then compares its answer with the one in the frame check sequence frame. It the two numbers match, the receiving station assumes it has received the frame without error. If the numbers do not match, the receiving station discards the frame and asks the transmitting station to retransmit. This operation is called a *cyclic redundancy check* or *CRC*.

End-of-frame detection.   The official frame ends at the end of the CRC frame. The receiver expects the line to be silent. However, it will sometimes still detect *dribble* bits, a last few bits that still occur after the end-of-the frame. The receiver can detect and simply discard up to eight dribble bits.

As Ethernet evolved, other frame formats also evolved. One example is the VLAN frame, which adds information about virtual networking (discussed later in this book). It adds 4 bytes of VLAN tag information, extending the maximum frame size from 1518 to 1522 bytes (Figure 4–3).

Preamble 7 bytes	Start Frame Delimiter 1 byte	Destination MAC Address 6 bytes	Source MAC Address 6 bytes	Length/ Type 802.1Q Tag Type 2 bytes	Tag Control Information 2 bytes	Length/ Type 2 bytes	Data 0–1500 bytes	Pad 0–46 bytes	Frame Check Sequence 4 bytes

FIGURE 4–3 Ten fields in VLAN frame

Interframe Gap

Each frame must be separated from another by a gap of silence. This permits a receiving station to better detect the end of a transmission. The interframe gap is 96 bit times, so the actual time in seconds is:

10 Mb/s Ethernet:	9.6 microseconds (9600 nanoseconds)
Fast Ethernet:	960 nanoseconds
Gigabit Ethernet:	96 nanoseconds

A Note on Transmission Rate and Data Rate

A point of confusion is the difference between the transmission rate of Ethernet and the rate at which you can expect your data to go flying around the network. The important thing to realize is that the speed of Ethernet is the speed at which Ethernet is spitting bits down the wire. As you can see from the Ethernet frame in Figure 4-2, only some of these bits are data. The rest are information on addresses, CRC, and so forth—important to the network operation, but overhead. Take the maximum Ethernet frame. It is 1518 bytes long and contains 1500 bytes of data. Over 98% of the frame is data. It has very little overhead. In the minimum case, the frame contains 46 bytes of data in a 64-byte frame. Only about 71% of the frame is actual data. Transmitting a full frame is very efficient when pumping data across the network. Very small frames are much less efficient since they contain a much higher percentage of overhead.

ACCESS METHOD

Any network requires a method for a station to gain access to the transmission medium. With a baseband transmission medium like Ethernet, two signals cannot be simultaneously transmitted on a line. The access method provides a reasonable method to arbitrate access to the medium and prevent two or more messages from being sent on a conductor at the same time.

There are several access methods available. For example, token ring networks use a special message called a *token* that passes from station to station around the network. Only when a station has possession of the token is it allowed to transmit. In the 1980s, the token ring network was an important competitor to Ethernet, but it fell to the wayside when it failed to keep pace in performance with Ethernet.

Another method is a master/slave arrangement. Here one station serves as a master and gives permission to other stations to transmit. This permission may be in the form of polling, where the master asks each station if it wants to transmit. Or stations wanting to transmit request permission from the master.

The common feature of token ring and master/slave networks is that a station can only transmit when it has permission, by acquiring the token or receiving authorization from the master.

Carrier Sense, Multiple Access with Collision Detection

The access method used by Ethernet is called **carrier sense, multiple access with collision detection** or **CSMA/CD** for short. The term sums up the main features of the access method:

Carrier sense.   Before transmitting, a station listens to the line to see if another station is transmitting. If it does not sense a transmission on the carrier (the line)—if the line is quiet— then it assumes it can transmit.

Multiple access.   Each station has an equal right to attempt to transmit, without waiting for permission.

Collision detection.   The possibility always exists that two stations will listen, sense a quiet line, and attempt to transmit. Such multiple transmissions are not allowed and are called **collisions.** If a collision is detected, both stations cease transmitting, wait a brief but random time, and begin the process over by listening to the line. When a station detects a collision, it transmits a special signal known as a *jam signal,* which notifies all stations that a collision has occurred.

Figure 4–4 shows a collision sequence. Station 2 listens to the network, detects no traffic, and begins to transmit. Meanwhile, Station 3 wants to transmit and listens. Because the signal transmitted by Station 2 has not arrived at Station 3, the station detects quiet and starts to transmit. A collision occurs when the two transmissions meet. Station 1 will detect the collision and issue a jam signal that will jam the entire network for a specific amount of time. All stations will receive the jam signal and know there has been a collision. When Stations 2 and 3 receive the jam signal, they will back off, wait a random period, then try again.

Some people treat a collision like an error or fault. It is not. It is simply the method used to arbitrate access to the network. While it is often desirable to reduce or eliminate collisions to enhance network performance, collisions themselves are not "mistakes" in the network's operation. Even so, industrial applications typically require low-collision or collison-free operation.

How Long Must I Wait?

We said that the stations involved in the collision will back off for a brief, random time before attempting to retransmit. How brief? How random? That depends on the size of the network and the number of consecutive collisions a station has experienced. Figure 4–5 shows the maximum backoff time for a 10-Mb/s Ethernet system.

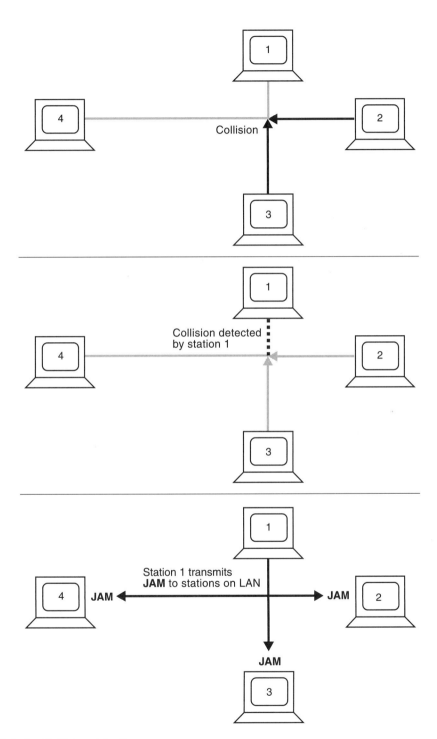

FIGURE 4–4   Collision detection sequence

Collision-on-attempt number	Maximum random number	Range of backoff times	
		10 Mb/s	100 Mb/s
1	1	0 ... 51.2 µs	0 ... 5.12 µs
2	3	0 ... 153.6 µs	0 ... 15.36 µs
3	7	0 ... 358.4 µs	0 ... 35.84 µs
4	15	0 ... 768 µs	0 ... 76.8 µs
5	31	0 ... 1.59 ms	0 ... 15.9 µs
6	63	0 ... 3.23 ms	0 ... 323 ms
7	127	0 ... 6.50 ms	0 ... 650 ms
8	155	0 ... 13.1 ms	0 ... 1.31 ms
9	511	0 ... 26.2 ms	0 ... 2.62 ms
10–15	1023	0 ... 52.4 ms	0 ... 5.24 ms
16	NA	Discard packet	Discard packet

FIGURE 4–5   Backoff times for a 10-Mb/s or 100-Mb/s network

When a station attempts to transmit and detects a collision, it first determines the range of random numbers. The random number must be an integer. As you can see from Figure 4–5, the range is from 0 to 1 after the first collision. The random number is then multiplied by the Ethernet slot time, which is 51.2 µs for a 10-Mb/s system or 5.12 µs for a 100-Mb/s system (we will describe the slot time and its additional significance in a moment).

For now, let us look at the 10-Mb/s example with a 51.2-µs time slot. The backoff time after the first collision is either 0 or 51.2 µs. Notice that each involved station will back off a different amount of time based on its calculation of the random number.

After waiting the calculated backoff time, a station will begin the media-access process over. If it again detects a collision, it will again calculate the backoff time. This time, however, the random number is between 0 and 3. The possible backoff time is zero, one, two, or three slot times: 0 µs, 51.2 µs, 102.4 µs, or 153.6 µs.

With each consecutive collision a station experiences, the range of random numbers and, therefore, the range of backoff times increases. For 10 to 15 collisions, no increase in the time occurs, although there are now 1024 (including 0) possible random numbers extending to 52.4 ms. At the sixteenth collision, the situation is deemed hopeless and the frame transmission is discarded.

CSMA/CD may seem a democratic method for giving all stations equal access to the network and resolving collisions. But it does give a slight edge to a station that successfully "wins" the first collision. Consider two stations, A and B, both of which attempt to transmit.

Suppose in calculating the backoff that station A is allowed to transmit immediately while station B has to wait 51.2 μs. Now, go further and suppose stations A and B again try to transmit only to have another collision. This time, station A has a decided statistical advantage. It still must calculate a backoff time of 0 μs or 51.2 μs. Station B, since this is collision number 2, will arrive at 0 μs, 51.2 μs, 102.4 μs, or 153.6 μs. The probability is that station A will be able to transmit first. If this continues with additional collision, station B will have a potentially longer wait each time, giving station A an advantage.

For station A, the best chances are:
    Collision, transmission, collision, transmission, collision . . .
This being so, station B is left with the worst-case scenario:
    Collision, collision, collision, collision

Remember, though, this is a random process and the possibility exists that station B will pick the lower number and win the right to transmit. Once station B wins the right to transmit, the fortunes are reversed, and it has the advantage.

THE IDEA OF DETERMINISM

Ethernet sometimes gets a bad rap because of the randomness of the CSMA/CD access method. Many network designers want to be able to determine with a reasonable degree of accuracy how long it will take from the time a station begins to transmit until a receiving station receives the message. Collisions throw randomness into the calculations. In industrial applications especially, predictability in transmission times can be important in keeping automation equipment or industrial processes running properly. Indeed, an important feature in the widespread adoption of industrial Ethernet is means to reduce or eliminate indeterminism. Will a transmission arrive in 25 μs or in 127.4 μs because of collisions?

There are six important things to remember about collisions and determinism. First, remember that collisions are not mistakes: they are the built-in method for arbitrating access to the network.

Second, depending on the application, even a large number of collisions may be tolerated. For example, in an office, the sending of e-mail will not even notice a high number of collisions. Users will not know if it takes a few extra milliseconds for an e-mail to arrive because of collisions.

Third, collisions can be minimized in many cases by careful network design. For example, you can limit the number of stations in the collision domain and limit the frequency and sizes of messages transmitted over the network. A lightly loaded network can operate with very few collisions.

Fourth, determinism is in the eye of the beholder. For example, a network could be built with a low incidence of collisions so that the longest backoff time is only 153.6 μs. Even in a factory application, where response times of several milliseconds are critical, this variation of up to 153.6 μs could be insignificant. From the automation engineer's viewpoint, this 153.6-μs window of variation is within the limits of required determinism. Even with collisions, the timing isn't considered indeterminate.

Fifth, collisions can be *eliminated* with certain versions of Ethernet. No collisions occur.

Six, it is sometimes erroneously assumed that simply adding more speed to the network will solve some of the determinism concerns. In a factory, it is the predictability of the timing—not the speed—that is critical. Simply increasing the transmission speed from 10 Mb/s to 100 Mb/s won't necessarily solve the problem. Certainly higher speeds are nice—signals get through the network faster so there's less likelihood of collisions (all things being equal).

Slot Time

The slot time is the basic unit in determining the backoff time. The time is related to the smallest packet that can be transmitted while ensuring that the transmitting station recognizes a collision and ceases transmission. The slot time varies depending on the speed of the network:

10-Mb/s Ethernet:	512 bits
Fast Ethernet:	512 bits
Gigabit Ethernet:	4096 bits

COLLISION DOMAIN

The **collision domain** (Figure 4–6) includes all attached stations that are subject to the collision rules. That is, it includes all stations that must listen for traffic from other attached stations and back off if a collision occurs. As we will discuss later, it is possible to create a net-

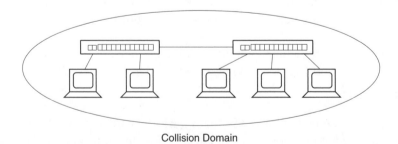

Collision Domain

FIGURE 4–6   Collision domain: All stations are subject to collision rules and can collide with all other attached stations

work having several different collision domains. By reducing the number of stations in a collision domain, you also reduce the traffic and the potential for collisions. A bridge, for example, is a simple device for dividing a network into separate collision domains. All stations on one side of the bridge share a collision domain. The stations on the other side of the bridge share a different domain.

UNICASTS, BROADCASTS, AND MULTICASTS

There are different types of Ethernet transmissions, namely one-to-one, (unicast), one-to-many (multicast), and one-to-all (broadcast). Briefly, **unicast** transmissions are between two stations: a transmitting station and a specifically addressed receiving station. At times, you want to communicate with more than one station.

In a **broadcast** transmission, the message goes to all stations within the broadcast domain. In a **multicast** transmission, the message goes to only selected stations within the broadcast domain. The Ethernet frame will have specific multicast or broadcast addresses in its destination address frame. Only one broadcast address is needed, since all stations are involved. Several multicast addresses may be used, since different types of multicasts can be used on the network.

One example of a broadcast transmission occurs when a station joins the network. It will broadcast its address to all other stations to let them know it is on the network. In return, each station will reply with a unicast transmission, which will allow the new station to know all the other stations on the network.

Broadcasts and multicasts add to the transmission efficiencies of Ethernet when the same information must be communicated to more than one station. Rather than send a separate, individually addressed message to each station, a single multicast or broadcast requires only a single transmission. In factory automation applications, multicasts can be very efficient at conserving bandwidth and maintaining determinism. For example, if you need to download code or otherwise communicate with several PLCs, a multicast transmission is far more efficient.

Figure 4–7 shows the differences between unicast, multicast, and broadcast transmissions. The consequences are detailed in Chapter 9.

ETHERNET AND THE OSI MODEL

The Open System Interconnection is a reference model issued by the ISO in the 1970s. Its purpose is to establish greater conformity in the design of networks by defining the network in a layered, modular fashion. It offers a seven-layer approach to describing a network and

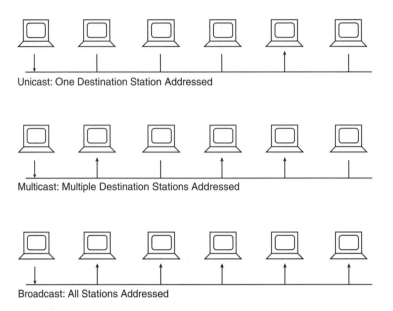

Unicast: One Destination Station Addressed

Multicast: Multiple Destination Stations Addressed

Broadcast: All Stations Addressed

FIGURE 4–7   Unicast, multicast, and broadcast transmissions

describes how the layers of the network interact and exchange information. The advantage of layers is simple: each layer provides certain functions necessary for the operation of a network. Each layer communicates with the one above it and the one below it. It reduces the complexity of the network from one large problem to seven smaller ones. Engineers need work on only one layer in a modular fashion, without worrying about other layers. It also helps ensure interoperability of technology, allowing a new layer to be developed with the confidence that it will work with adjacent layers. Ethernet, as you will see, is concerned with the lower two layers. It can work with numerous upper layer designs.

It is important to remember that the OSI is a reference model. That is, it specifies the functions of the layers, but it does not provide a real-life example. It is only a framework for describing the network. An actual network does not have to conform to these standards, nor does it have to have seven layers. Even so, the ISO model is extremely convenient for conceptualizing a network.

Figure 4–8 shows the OSI model. The upper four layers are the host layers, responsible for accurate delivery between computers. The lower three layers are the media layers whose function is to see that the information arrives at the proper destination. The middle layer, the transport layer, provides the interface between the upper and lower layers. To a large degree, the host layers perform functions closer to what you as a user want. The lower three layers are concerned with getting information onto and off of the network reliably.

The applications you use on a computer—such as a word processor, spreadsheet, or presentation program—exist above the application layer. The application layer provides network services to the applications. Notice that it is the application layer that provides the services,

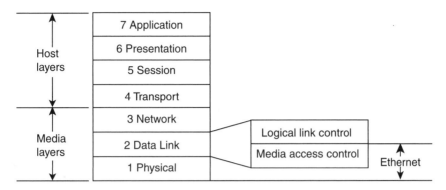

FIGURE 4–8   OSI model

not another layer. The OSI model is hierarchical. Layers only perform services for those above or below.

Here is a brief description of the functions of each layer.

Physical Layer

The physical layer is the most basic, concerned with getting data from one point to another. This layer includes the basic electrical and mechanical aspects of interfacing to the transmission medium. This includes the cable, connectors, transmitters, receivers, and signaling techniques. Thus it defines the electrical (or optical) characteristics of the signal, the signaling speed, the type of cable used, and so forth. Each specific type (or subtype) of network defines the physical layer differently. Thus, 10BASE-T (using twisted-pair cable at 10 Mb/s), 100BASE-TX (using twisted-pair cable at 100 Mb/s), and 100BASE-F (fiber at 100 Mb/s) all require a different physical layer implementation.

Data-Link Layer

The data-link layer provides for reliable transfer of data across the physical link. This layer establishes the protocols or rules for transferring data across the physical layer. It puts strings of characters together into messages according to specific rules and manages access to and from the physical link. It ensures the proper sequence of transmitted data.

Networks like Ethernet and Token Ring exist on the physical and data-link layers. In a PC network, chips on the PC's network interface card perform the physical and data-link functions. Software in the computer performs the functions of the higher layers. From the viewpoint of this book, building cabling and networks are distinguished at the physical and data-link layers.

In Ethernet, the data-link layer is subdivided into two parts: the Ethernet media access control (MAC) sublayer and the logical link control (LLC) sublayer. The LLC is defined in IEEE 802.2.

The MAC, as its name implies, controls access to the media—the physical layer and attached cable. It is the MAC that uses CSMA/CD to manage access. It listens to the network, detects collisions, and performs the backoff algorithm. In most respects, the MAC is the heart of Ethernet.

Network Layer

The network control layer addresses messages to determine their correct destination, routes messages, and controls the flow of messages between nodes. While there are several popular network layer protocols, the one rapidly gaining the most widespread acceptance, particularly in factory networking, is the Internet protocol (IP).

IP is a routing protocol, meaning that it contains both the address of a network *and* the address of a station within that network. The Ethernet MAC address, in contrast, assumes that all addresses are within the same network. While IP is the addressing scheme of the Internet, connecting people all around the world into the World Wide Web, it can be used within smaller networks.

A station's IP address can be permanently assigned or it can be assigned at startup. For example, an office computer connected to the Internet over a corporate network typically has a fixed, permanent address assigned. If you use dial-up service from you home, your Internet service provider typically assigns you an IP address each time you dial in.

Transport Layer

This layer provides end-to-end control once the transmission path is established. It allows exchange of information independent of the systems communicating or their location in the network. The two most popular protocols used in industrial networking are transmission control protocol (TCP) and user datagram protocol (UDP).

TCP is defined as a connection-oriented protocol whose task is to provide reliable data transmission between devices. Connection-oriented protocols require that a connection first be established between transmitter and receiver so that data can pass back and forth. The most obvious example of a connection-oriented network is the telephone network: You cannot communicate unless a connection is made at both ends.

TCP handles the fragmentation and reassembly of messages, detects failures, and performs retransmission. Because it monitors the reliable delivery of messages, TCP ensures a high quality of services in guaranteeing delivery. If transmission fails, TCP notifies both stations.

TCP breaks a large message into smaller segments for transmission. In some cases, these different segments can be routed over different paths so that they arrive at the receiving station in a different order than that in which they were transmitted. TCP at the receiving end must be able to reassemble the segments in the correct order. As TCP breaks the message into segments, it numbers each segment so they can be correctly sequenced by the receiving station.

The quality control performed by TCP means that it is not the fastest protocol. It is useful in factory application for downloading programs between a computer and a PLC, for operator control stations that read and write to PLC data tables, and for passing messages between PLCs. For real-time communications with I/O and similar needs, UDP is the preferred protocol.

UDP is a connectionless protocol, much simpler than TCP. Its purpose is to send datagrams between two devices. Because it is simpler, UDP is faster, able to satisfy the real-time requirements of applications. The penalty is the lack of guaranteed delivery at the transport layer. This does not mean that the protocol is unreliable. First, handshaking (or the checking of connections between stations) can be done by applications. Second, delivery of datagrams *is* usually reliable.

Session Layer

The session-control layer controls system-dependent aspects of communication between specific nodes. It allows two applications to communicate across the network.

Presentation Layer

At the presentation layer, the effects of the layer begin to be apparent to the network user. This layer, for example, translates encoded data into a form for display on the computer screen or for printing. In other words, it formats data and converts characters. For example, most computers use the American Standard Code for Information Interchange (ASCII) format to represent characters. Some IBM equipment uses a different format, the Extended Binary Coded Decimal Information code (EBCDIC). The presentation layer performs the translation between these two formats. (Microsoft Windows performs presentation layer functions.)

Application Layer

At the top of the OSI model is the application layer, which provides services directly to the user. Examples include resource sharing of printer or storage devices, network management, and file transfers. Electronic mail is a common application-layer program. This is the layer directly controlled by the user.

In practice, networks work from the top layer of one station (the message originator) to the top of another station (the message recipient). A message, such as electronic mail, is created in the top presentation layer of one workstation. The message works its way down through the layer until it is placed on the transmission medium by the physical layer. At the other end, the message is received by the physical layer and travels upward to the presentation level. It is at the presentation level that the electronic mail is read.

In factory applications, the top three layers (session, presentation, and application) are sometimes lumped together and referred to as the application layer. Remember that the OSI model is not mandatory and you do not have to divide everything into layers. The reason for lumping the three layers together for our purposes it that the application will perform all three functions even if viewed as one layer, or two, or three.

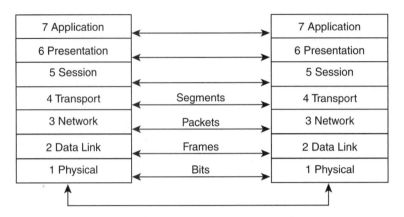

7 Application	←──────────────→	7 Application
6 Presentation	←──────────────→	6 Presentation
5 Session	←──────────────→	5 Session
4 Transport	Segments ←───→	4 Transport
3 Network	Packets ←───→	3 Network
2 Data Link	Frames ←───→	2 Data Link
1 Physical	Bits ←───→	1 Physical

FIGURE 4–9   The relationships between bits, frames, packets, and sessions

As discussed, Ethernet proper involves only one and a half layers—the physical layer or PHY and the bottom part of the data-link layer. The top half of the data link is the link logic control, which is defined by IEEE 802.2. The part of the data-link layer defined by Ethernet is the media access control (MAC) sublayer. In other words, Ethernet defines the PHY and MAC.

Ethernet can operate with any number of upper layer protocols. While TCP/IP is becoming the overwhelming favorite because it is the protocol suite of the Internet, other protocols work equally well. In particular, though, TCP/IP is becoming a universal protocol, replacing existing protocols. But we do not wish to put too sharp a point on the growing popularity of TCP. Existing protocols still have a place in many applications, particularly existing applications where the costs and difficulty of changing make it prohibitive.

Review of Layers, Encapsulation, Bits, Frames, Packets, and Sessions

Within the OSI model, intelligence is added at each step up the model. Working from the bottom up, the true information—the message, what you or a machine needs to communicate—spits out of the application layer. At the other end, the physical layer exists to swallow binary bits. At this level the bits are represented as electrical pulses. At the next level, the MAC assembles groups of bits into an Ethernet frame, with addressing and other information wrapped around the data. An Ethernet frame, of course, contains information for the network.

As shown in Chapter 1, IP adds a header containing addressing information, and the unit is now functioning as a packet. TCP again adds a header, and the unit now functions as a session. Figure 4–9 shows the functional relationship between these units.

In communicating, each level handles its own type of unit. Transmitting and receiving physical layers are concerned only with bits. Similarly, the MAC layer looks to the Ethernet frame for information, the network layer looks for the IP header information in a packet, and the transport layer looks for TCP information in the session.

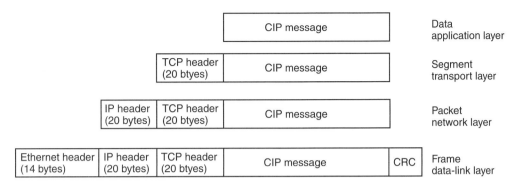

FIGURE 4–10   Encapsulation

The encapsulation process that wraps bits into frames and frames into packets is just as important for industrial Ethernet protocols as it is for Internet protocols. Many high-level automation data and protocols are encapsulated into TCP/IP for transmission over Ethernet. Figure 4–10 shows the idea of encapsulation for an iE protocol known as Ethernet/IP.

INDUSTRIAL ETHERNET: THE DEFINITION EXPANDED

When most people talk about industrial Ethernet, they mean more than the PHY and MAC layers defined by 802.3. For industrial applications, the top three layers are equally important in achieving a successful Ethernet-based factory network. Simply put, industrial applications require application-level protocols that will define the high-level communications between equipment. These protocols will allow equipment to talk to one another, a consideration that is increasingly important in achieving open systems. If an Ethernet-enabled I/O from vendor A is to talk to a PLC from vendor B, then you need a high-level application that each understands.

TCP, for example, guarantees a message will be delivered from station A to station B. It makes no such guarantee that the message will be understood. The message may be delivered, but communication doesn't necessarily happen. The application-layer software must define the common grounds of communication.

As of this writing, there are many efforts to define these applications:

- **Modbus/TCP,** which bundles the Modbus application protocol on top of TCP.
- **Interface for Distributed Automation (iDA),** which is a publisher/subscriber model using Network Data Delivery Services (NDDS) as middleware.
- **Foundation Fieldbus High-Speed Ethernet (HSE),** which maps fieldbus layers onto Ethernet and TCP/IP.

- **EtherNet/IP,** which uses the existing control and information protocol (CIP) layer of DeviceNet and ControlNet over TCP/IP and Ethernet.
- **ProfiNet** uses Microsoft's Distributed Component Object Model (DCOM) over TCP.

These and other iE protocols are examined in Chapter 10.

All of these approaches have support from major players in the automation industry. Whether one or two will emerge as overwhelming winners is anybody's guess. However, most companies recognize that repeating the fieldbus wars benefits few people and that a single, unified industrial Ethernet standard offers the widest benefits. Still, providers of automation equipment have turf to protect and do not always want to play nice with competitors. While each approach is billed as an open standard (and each is to a large degree), users benefit most when there is a single open standard, not five or more.

These application-layers are important to widespread adoption of industrial Ethernet. In many respects, these layers remove a certain level of indeterminism from the systems. The application layers do several important things for an Ethernet-based system:

- They define the data structures, so that sizes of data streams are easier to predict. This allows more predictable Ethernet frame sizes and more predictable network loading.
- They define reliability and error correction to allow use of more efficient UDP when necessary.
- They establish sophisticated and efficient mechanisms, such as object models and publish/subscribe protocols, to make network operation more efficient.

SUMMARY

- Ethernet is a network with many variations in speed, cabling, and topology.
- The variations most appropriate to the factory floor are 10BASE-T and 100BASE-T (Fast Ethernet).
- The Ethernet frame is of variable length and can carry from 48 to 1500 bytes of data.
- The frame also includes addressing information for both the source and destination and error control information.
- Ethernet uses a media access technology known as carrier sense multiple access with collision detection (CSMA/CD).
- CSMA/CD is the main source of indeterminism, caused by the random backoff of stations after a collision.
- The OSI model is a seven-layer model describing a network.
- There are many different protocols used to construct a layered network model.
- Ethernet defines the lower physical layer and half (the MAC sublayer) of the data-link layer.
- Industrial Ethernet includes Ethernet, TCP/IP for the network and transport layers, and one of several application protocols for the upper layers.

? REVIEW QUESTIONS

1. What three companies developed the first version of Ethernet?
2. What IEEE committee controls the Ethernet specification?
3. What was the data rate of the original version of Ethernet?
4. What are the five rates that Ethernet operates at today?
5. What are the minimum and maximum number of data bytes in an Ethernet frame?
6. How many layers does the standard OSI model have? What layers does Ethernet operate on?
7. What is the name of the access method used by Ethernet?
8. What is the main way Ethernet's access method causes indeterminism?
9. What is the difference between a packet and a frame?
10. If a 10-Mb/s station encounters three successive collisions, what is the maximum backoff time it must wait before attempting to transmit again?

A Closer Look at Ethernet

5

OBJECTIVES

After reading this chapter, you will be able to:

- List the three types of Ethernet cable
- Define the main flavors of Ethernet in terms of cable and data rate
- List eight differences between commercial and industrial requirements for Ethernet equipment
- Describe basic differences among hubs, switches, bridges, and routers
- Describe how determinism is achieved in a full-duplex switched network

FLAVORS OF ETHERNET

As we have mentioned, Ethernet has grown and evolved over the years into many variations or "flavors." This section looks at the different variations, with emphasis on those that are most applicable to industrial networking.

The main distinction between the different flavors of Ethernet lies in the data rate and the type of cable used. The speed is straightforward: Ethernet runs at the following speeds:

- 10 Mb/s
- 100 Mb/s (Fast Ethernet)
- 1 Gb/s (1000 Mb/s) (Gigabit Ethernet)
- 10 Gb/s (10,000 Mb/s) (10 Gigabit Ethernet)

Of these, 10 Mb/s and 100 Mb/s Ethernet are the most widely used in industrial applications. Fast Ethernet, in particular, deserves attention since it is fairly inexpensive and offers performance well matched to demanding applications. Gigabit Ethernet today is mostly used to connect to servers and for backbone applications. It is slowly migrating toward the desktop and may eventually find application on the factory floor. Ten Gigabit Ethernet is new and is expected to find application in metropolitan and very-high-bandwidth corporate applications.

Three main types of cables, each with several subtypes, are used with Ethernet. For now, we will quickly introduce them.

- *Coaxial cable* was the type of cable originally used with Ethernet. All but obsolete in office applications because of its size, its stiff, hard-to-work-with construction, and its higher costs, coaxial cable still finds use on the factory floor. Its main attraction is its excellent immunity to electrical noise generated by factory equipment.
- *Twisted-pair cable* is small, flexible, and inexpensive. Similar to the cable used for telephone connections, twisted-pair cable uses pairs of wires twisted together. There are several grades or categories of this cable (running from lower performance to higher performance): categories 3, 5, 5e (e for enhanced), and 6. One other emerging type is category 5i for industrial applications. In general, category 3 is suited to 10 Mb/s applications and category 5/5e/5i cables are the choice for factory applications at rates of 10 Mb/s, 100 Mb/s, and even 1 Gb/s. Category 6 is even better, but for reasons that will be explained later, it is not as well suited to the electrical environment of some factories.

  Twisted-pair cable allows runs of up to 100 meters. It is the workhorse cable of Ethernet networks.
- *Fiber-optic cable* uses glass strands to carry signals as light. The optical fiber is the best transmission medium of the three types of cable: It can carry higher data rates over longer distances. While twisted-pair cable is generally limited to 100-meter runs, fiber-optic cables can transmit thousands of meters. Equally important, optical fibers are immune to the electrical noise generated in the factory.

  The drawback to fiber is that it is perceived to be fragile (which do you think is sturdier: glass or copper?). However, fiber-optic cables are quite rugged and can meet application demand. The biggest hindrance to widespread use of fiber is its higher costs (perceived or real).

There are several variations of fiber-optic systems used in Ethernet, which relate to the type of fiber used and the type of light transmitted through the fiber. Here are the distinctions in a nutshell:

Multimode fiber can carry signals several hundred meters. *Single-mode* fiber can carry signals several thousand meters.

Short-wavelength light is less expensive, but not as well matched to the fiber to achieve the high transmission efficiencies. Long-wavelength light is better, but costs more.

Here, then, are the flavors of Ethernet.

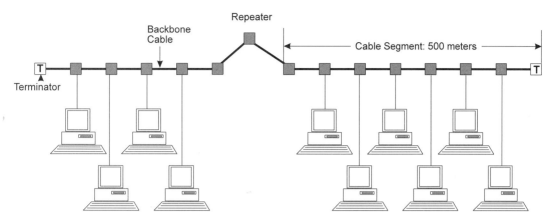

FIGURE 5–1 10BASE-5 network

10BASE-5

10BASE-5 is the original version of Ethernet, running at 10 Mb/s over a thick coaxial cable. Because of the cable, it is sometimes called *thicknet*. 10BASE-5 allows backbone lengths of 500 meters for each segment. Nobody uses 10BASE-5 today. Figure 5–1 shows a 10BASE-5 network.

10BASE-2

10BASE-2 is thin Ethernet, using a smaller, more flexible coaxial cable. Each segment can be 185 meters long, with up to five segments per network. The distance from the backbone to a station is 50 meters maximum. Those applications using coaxial cable are 10BASE-2. See Figure 5–2.

1BASE-T

1Base-T, often called StarLAN after an AT&T version of the network, was the first to run over telephone cable—even the cheaper stuff not covered by the categories of twisted-pair cable. Originally designed as an inexpensive alternative to coaxial cable-based networks, it never succeeded. 10BASE-T has more to offer.

10BASE-T

10BASE-T is Ethernet on twisted-pair cable. 10BASE-T departs from the bus structure of 10BASE-5 and -2 to use a star-wired configuration. All stations are connected through a center point (a hub or switch). The maximum distance from the hub to the end station is 100 meters. Figure 5–3 shows a 10BASE-T network.

10BASE-T networks can operate on the category 3, 5/5e/5i, and 6 cable described later in Chapter 7—in other words, all common twisted-pair cable. The maximum recommended distance from switch or hub to station is 100 meters. While distances of 150 meters can be

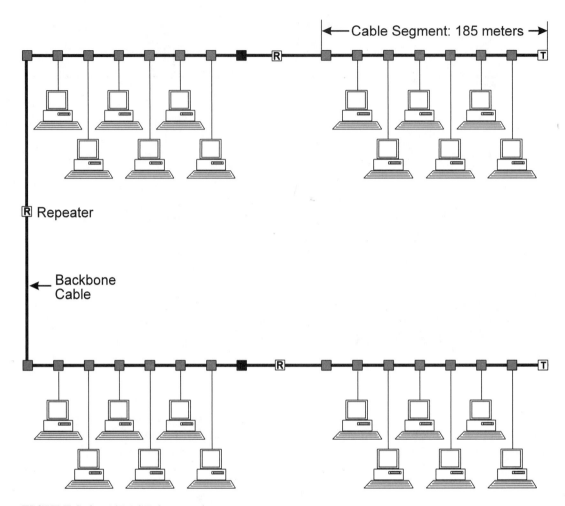

FIGURE 5–2 10BASE-2 network

achieved with category 5 cable, exceeding 100 meters is not recommended. Since the cable distance is from station to hub or switch, two stations can be as far apart as 200 meters when they are connected to the same hub or switch. Hubs and switches, too, can be interconnected to extend the distance between two stations. Again, the recommended distance between two hubs or switches connected over UTP is 100 meters. Many 10BASE-T devices also allow fiber backbones that extend the distance between devices to 2 km. You can see that the 100-meter station-to-hub/switch limit is really flexible and not so limiting as it first appears.

10BASE-F

The 10BASE-F specification adds a fiber-optic alternative for use in IEEE 802.3 networks. The specification actually lists three different variations: a backbone cable (10BASE-FB) to

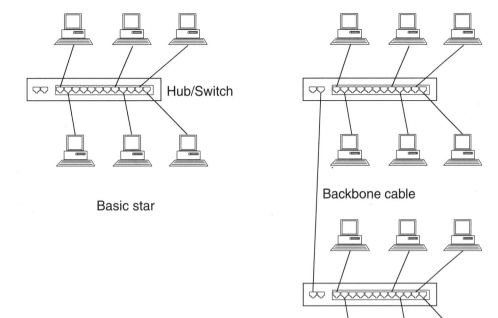

Hub/Switch

Basic star

Backbone cable

Two stars connected

FIGURE 5–3   Star-wired 10BASE-T network

connect hubs/switches, a passive star-coupled network (10BASE-FP) that splits light from one fiber into many, and a fiber-optic link between hub and station (10BASE-FL). Most applications use 10BASE-FL, a simple point-to-point link between two devices. The requirements for a 10BASE-FB and a 10BASE-FL are fairly straightforward. The backbone is simply a point-to-point connection between hubs. The FL version defines connection between a hub and station or between hubs or switches. Each point-to-point link can be up to 2 km long.

Fast Ethernet

Fast Ethernet brings a 100-Mb/s upgrade path to Ethernet networks. Also known as *100BASE-X,* Fast Ethernet uses the same frame format and medium access-control mechanism as its 10-Mb/s sibling. It allows the same applications and network software, but at a tenfold increase in transmission speed.

Currently, Fast Ethernet recognizes four variations:

- 100BASE-TX is probably the most popular flavor of Ethernet today. It operates at 100 Mb/s over category 5 cable. To maintain compatibility with 10BASE-T, it uses the same wiring assignments in the connectors. Beginning with Fast Ethernet, the signaling techniques for copper cable are more complex than the Manchester codes described earlier. The purpose of the complex codes is to allow a high data rate to be sent over a cable at a lower frequency.
- 100BASE-FX defines a fiber-optic link segment.
- 100BASE-T4 uses four-pair cable to allow transmission over category 3 cable. The standard recommends category 5 connecting hardware for connectors, patch panels, punchdown blocks, and so forth. Of the four pairs, one transmits, one receives, and two are bidirectional data pairs. Because category 5 cable has proven so popular, 100BASE-T4 is not widely applied. It was originally specified to allow the existing installed base of category 3 cable to handle Fast Ethernet.

100BASE-T defines two classes of hubs. Class I hubs can connect unlike media so that segment types can be mixed. A class I hub can have ports accepting two-pair UTP (100BASE-TX), four-pair UTP (100BASE-T4), and fiber. (Remember that, although 100BASE-TX only uses two pairs for signaling, it is usually wired with four-pair cable.) A class II hub accepts only one media type.

Fast Ethernet allows 100-meter cable runs. In this case, the limit is set by timing considerations for the round-trip delay of a signal. It is built into the specification. In contrast, the 100-meter limit recommended for 10BASE-T is artificial in the sense that it is not a physical limitation of the network. With category 5 cable, 10BASE-T can easily support 150-meter cable runs, although 100 meters is the generally accepted maximum.

Gigabit Ethernet

Gigabit Ethernet, operating a 1000 Mb/s, is even faster—ten times faster than Fast Ethernet. The main application is to connect switches and hubs to one another and to network servers. Because Gigabit Ethernet ports tend to be expensive, they are found in workstations only for the most demanding applications, such as 3D modeling. During 2002, prices began dropping significantly.

Four flavors of Gigabit Ethernet are available:

- *1000BASE-TX*. This is Gigabit Ethernet over UTP. While most network applications use two pairs (one for transmitting and one for receiving), Gigabit Ethernet uses all four pairs to transmit in both directions and can transmit and receive simultaneously over a cable. In essence, the 1000-Mb/s signal is broken into four 250-Mb/s data streams. Each stream is then sent over a different pair and then re-assembled by the receiver.

It is this use of all four pairs that allows Gigabit Ethernet to operate over category 5 cable. By splitting the signal and then using sophisticated PAM5 encoding, the frequency of the signal falls under 100 MHz. The original specification of category 5 did not consider the demands that Gigabit Ethernet would place on the cable. Most experts recommend category 5e as a better choice for Gigabit Ethernet.

- *1000BASE-CX.* This flavor allows Gigabit Ethernet over short links of no more than 25 meters using shielded cable, including coaxial cable. The main purpose is for short connections between equipment in the same wiring closet.
- *1000BASE-SX.* This covers multimode fiber-based systems using a short wavelength, typically 850 nm.
- *1000BASE-LX.* This flavor is Gigabit Ethernet over multimode or single-mode fibers using a long wavelength in the 1300-nm region. This longer wavelength offers better performance: At this wavelength, fibers have higher bandwidth and lower attenuation than at shorter wavelengths. But 1300-nm transceivers tend to be more expensive.

Most networks use a combination of Ethernet speeds. Gigabit Ethernet is used mainly for connections to servers and for backbone connections between hubs. Connections to the desktop are typically with Fast Ethernet in newer applications and 10-Mb/s Ethernet over older hubs and switches. Most newer equipment—hubs, switches, and NICs—have autosensing ports that will adjust the speed between Fast Ethernet and 10-Mb/s Ethernet, depending on the speed of the port at each end. In short, Fast Ethernet is fast becoming the general-purpose solution for most Ethernet applications. Still, the widespread base of 10-Mb/s Ethernet means that there are still many computers and other devices running at this slower speed.

10-Gigabit Ethernet

Ethernet operating at 10 Gb/s is the latest variation. However, this version has little applicability to the factory floor. While businesses may adopt it for specific needs requiring very high bandwidth, 10-Gigabit Ethernet is viewed as a solution for metropolitan applications or for very-high-speed backbones within a company. While a full understanding of 10GBASE-X variations is not required for factory networking, we give a brief description.

10-Gigabit Ethernet is divided into two main variations: local area networks (LANs) and wide area networks (WANs). The network uses optical fibers, since its high speed makes copper cable impractical except for very short runs of only a few meters. The systems operate in one of two main ways. The first uses a stream of light at 10 Gb/s. The second uses a technique called *wavelength-division multiplexing,* which divides the transmission into four separate 2.5-Gb/s streams and sends them simultaneously over the fiber. The 10-Gigabit Ethernet standard takes advantage of transmission techniques borrowed from existing standards (such as SONET, which covers optical telecommunications). One reason that Ethernet has evolved quickly is that it has adopted existing standards that have already solved certain technical hurdles.

ETHERNET ON THE FACTORY FLOOR

With all the choices of Ethernet, which is best on the factory floor? In general, Fast Ethernet is the all-around best choice, offering high data rates, attractive pricing, and a wide range of devices. 10 Mb/s, particularly 10BASE-T, can meet many applications economically, but the cost savings over Fast Ethernet are not dramatic enough to make the slower speed the choice. While 10BASE-T is by no means obsolete, it is fading while Fast Ethernet is growing in popularity. The choice is not either/or. You can mix Ethernet and Fast Ethernet in industrial applications.

Gigabit Ethernet, in most cases, is not required at the fieldbus or device level. It may find applications at higher supervisory levels. The price for Gigabit Ethernet is dropping. Using Gigabit Ethernet for server and backbone connections may indeed make sense.

ETHERNET DEVICES

In creating an Ethernet network, there are a few basic devices, from the Ethernet interface on the equipment to the hubs and switches that tie it all together. We are interested in four devices in particular:

- Ethernet station interface
- Repeater
- Hub (multiport repeater)
- Bridge
- Router
- Switch

Ethernet Station Interface

Any equipment connected to an Ethernet network must have an Ethernet port that the network cable plugs into. These are often called *network interface cards* (NICs) because they were an add-in card in early computers. Today, the Ethernet interface can be built directly into the equipment using as little as one main chip. In industrial PLCs, the interface is built into a plug-in module, as shown in Figure 5–4. While the interface can operate at 10 Mb/s, 100 Mb/s, or 1000 Mb/s, a common form today is a dual-speed interface operating at either 10 Mb/s or 100 Mb/s. When connected devices first begin communications, they first perform a handshake that will tell each side at what speeds they can operate. Usually, they will settle on the highest speed.

Some industrial equipment is limited to the speed it can run at. Some older equipment "upgraded" to include an Ethernet add-on port cannot handle Fast Ethernet data rates. In internal processing power, the equipment simply does not have enough power. It must operate at the

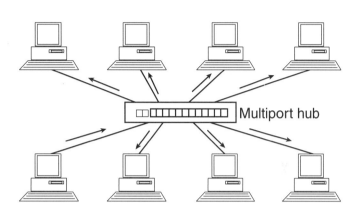

Multiport hub

FIGURE 5–4
Ethernet module
for a PLC
(Courtesy of Schneider
Electric)

FIGURE 5–5   A hub is a multiport repeater, forming a single
collision domain.

slower 10 Mb/s. Other equipment uses the 100-Mb/s rates as the standard data rate, so that slower
Ethernet is only minimally supported.

Repeaters

A repeater is a simple device that takes an incoming Ethernet signal and amplifies and retimes
it. A repeater is used to extend the physical size of a network. As distances increase, a signal
becomes degraded by two mechanisms: attenuation and skew. Attenuation is the loss of signal strength. Skew is the change of the signal from its perfect position in time. Think of skew
as a drummer who over time loses the perfect beat, going either too fast or too slow in relation to the ideal time of a metronome. Too much skew prevents the receiving circuit from
accurately determining if the received pulse is a 1 or a 0. A repeater corrects the problems of
attenuation and skew.

Hubs (Multiport Repeaters)

A hub is a multiport repeater. The device has several ports. An incoming signal is rebuilt and
then transmitted on all the remaining ports. The hub does not make any decisions about the
Ethernet frame or its contents. As shown in Figure 5–5, it simply relays it out all other ports.

The hub was the basic device that helped ensure Ethernet's popularity. In particular, hubs had three main features:

1. *Reliability.* They enabled a star-wired network. Any failure on any of the legs of the star did not bring down the entire network.
2. *Lower costs.* Hubs occurred as part of 10BASE-T; that is, they helped enable a low-cost, easy-to-implement alternative to coaxial cable. (Hubs are also available for coaxial networks, but do not offer the same economies.)
3. *Manageability.* Hubs are manageable. So-called smart hubs can be controlled and can monitor and report on their operation. Management of networks will be covered in Chapter 7. Dumb hubs, on the other hand, are extremely inexpensive, but are simple, mindless repeaters.

The drawback to a hub from the perspective of network performance in general and industrial networking in particular is its lack of intelligence. It is a shared media device subject to collisions and the resulting indeterminism in operation. We will return to this issue after we have reviewed other devices.

Bridges

As a network grows and more stations are attached, performance may decline as more and more devices vie for media access. A bridge allows the collision domain to be reduced by breaking a network into two sections. Consider the situation in Figure 5–6a. The single collision domain includes five stations connected through a hub. All five stations must share access. Figure 5–6b shows the same network connected by using a bridge. The network now contains two collision domains separated by a bridge.

The bridge separates the two networks by reading the MAC address of each Ethernet frame. If a message from a station in domain A is addressed to another station in domain A, the bridge does not pass the message on to domain B. Only when a message from domain A is addressed to a station in domain B does the bridge pass the message to domain B.

How does the bridge know which domain an address belongs in? There are two basic ways. The first is for an operator to laboriously enter a set of addresses into the bridge, along with an indication of their domain. This method is tedious and prone to error. An easier way is to allow the bridge to learn the addresses. Every time it receives a message, it also checks the source address. If it is a new address, the bridge enters it into an address table. Eventually, as all stations send messages, the bridge will learn the location of all stations.

If the station does not recognize the destination address, it will pass the message so that it goes to all collision domains. Equally important, the bridge needs some built-in mechanism for aging addresses. If an address neither receives nor sends messages over a sufficiently long time, it is removed from the table.

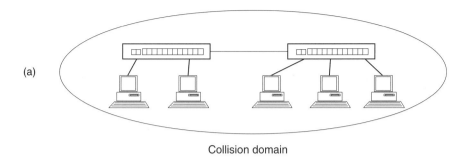

(a)

Collision domain

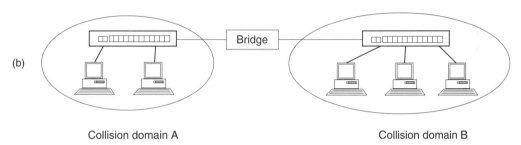

(b)

Collision domain A   Collision domain B

FIGURE 5–6   A bridge separates collision domains. It can be used to segment a network to improve performance.

Router

A router is similar to a bridge, except that it operates at the network layer, making the forwarding decision based on the IP address rather than the MAC address. The IP and subnet addressing schemes relevant to routers are detailed in Appendix B. This gives a router some additional capabilities: A bridge essentially "routes" information among addresses on a single network. A router contains information on both networks and individual addresses in a network. A bridge can control messages within a factory network, for example. A router can do this—plus it can decide which messages need to be sent to different networks the world over.

The Internet is built on routers to control traffic between all stations connected to the Internet.

Switches

You can think of a switch as a multiport bridge or router. Remember that a hub distributes the signal to *all* attached stations. It is a shared-media device that makes messages subject to collisions. A switch directs packets only to the station or stations for which it is intended. In other words, it *switches* the signal between the input port and the output port. Other stations do not receive the message. Figure 5–7 shows the operation of a switch.

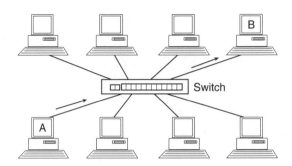

FIGURE 5–7   Switch operation

Ethernet equipment can work in either a half-duplex mode or full-duplex mode. In the half-duplex mode, the port cannot receive and transmit at the same time. It can transmit *or* it can receive. If it tries to do both, the received signal will be detected as a collision. In full-duplex, the port *can* transmit and receive at the same time. This effectively doubles the bandwidth of the network: If a station transmits at 100 Mb/s and receives at 100 Mb/s, effectively it is operating at 200 Mb/s. Full-duplex operation, when combined with switched-media operation, has an important consequence for industrial networks:

> *Full-duplex switched networks are not subject to collisions and, therefore, offer deterministic operation.*

A station operating in the full-duplex mode effectively shuts off its collision detection capability. It must, because a signal on the receive circuit while the station is transmitting is seen as a collision. Full-duplex operation requires that ports do not even look for collisions. Nor do they have to listen to the network before transmitting.

This is a very strange concept indeed! After all, Ethernet is by definition a network based on collision detection and carrier sense for media access. Remember the name of the 802.3 standard is *Carrier Sense Multiple Access with Collision Detection Access Method and Physical Layer Specifications.* Ethernet is not even mentioned. Now, with full-duplex switched operation, CSMA/CD is not required.

The reason CSMA/CD is not needed is that stations no longer have to contend for access to the media. Consider three stations, A, B, and C, connected together through a switch. If station A sends a message to station B, the switch will read the address information and direct the message only to station B. Nothing is sent to station C. At the same time, station C can send a message to station A. Again, the switch reads the destination address of station A and sends the message on. Full-duplex operation means that the message from station A and the message to station A can pass each other without ill effect.

So what happens if both station A and station C attempt to send a message to station B at the same time? Ideally, the switch will have enough buffer memory to store the second message (say station C's) while it sends station A's message. Otherwise, the switch will have to deal with one of the messages as lost and inform the sending station so it can resend. These types of situations are rare (and can occur with any type of network). The point remains that a full-duplex network does not have collisions.

The need to route is becoming more important even in LANs. Just a few years ago, conventional wisdom (based on empirical studies) showed that 80 percent of the data transmitted in a subnetwork stayed within that subnetwork. Only 20 percent was passed to another subnetwork. Today this traffic pattern is reversed. Only 20 percent of the traffic stays within the subnet. Over 80 percent travels beyond it. Among the many reasons are these:

- E-mail
- Internet use
- Consolidation of servers into easily managed server farms, which may be located on another subnetwork
- Users accessing multiple servers (e-mail server, application server, Internet server, print server . . .), which may be consolidated in a farm or spread out throughout the organization.

Fast, powerful processing chips and low-cost, high-speed memory enable switches. A hub does not have to ascertain anything about a message. It just repeats it out of every port. A switch must have the intelligence to dissect the message sufficiently to determine the destination address. Then it must look up to which port the destination station is attached. It then switches the message to the proper port.

CUT-THROUGH AND STORE-AND-FORWARD SWITCHES

There are two types of switching mechanisms used in switches. The first type used was the store-and-forward switch. Here the switch receives and stores the entire frame before sending it to the destination port. Even while it is receiving the full frame, it can be at work determining the destination.

Eventually, some manufacturers came out with a different switching mechanism that they claimed offered superior performance. These are the cut-through switches, which will begin sending the message to its destination as soon as the proper destination address is determined. Promoters of this technology claimed that cut-through switches were more efficient because they did not experience the delay of store-and-forward switches.

However, studies have shown that in terms of real-world efficiency, little difference exists between the two types of switches. While the cut-through switch does have an advantage with long frames, it also makes things more difficult when errors occur. Suppose the switch finds that the frame is invalid—a bad CRC, for example. The store-and-forward switch can drop the message and ask the sending station to retransmit. The cut-through switch has already begun forwarding the frame before it finds out there is an error.

It is not important which switch mechanism is used. The important measure is latency: How long does it take from the time the message begins arriving until it is done leaving. The latency will depend on several additional factors, including the network speed and the efficiency of the switch electronics. The best switches operate at "wire speed," meaning that they cause minimal delay to the frame. The switch can be treated as a wire of given length.

LAYER 2 AND LAYER 3 SWITCHES

Think back to bridges and routers. The bridge operates on layer 2 MAC addresses. The router uses layer 3 IP addresses. Now you know the difference between a layer 2 switch and layer 3 switch: The layer 2 switch is essentially a multiport bridge and the layer 3 switch is a multiport router. It is also called a *routing switch.*

You will see some manufacturers talk about layer 4 or layer 5 switches. Such talk, however, is more marketing hype than useful description. In a TCP/IP Ethernet network, address information exists on layers 2 and 3. Higher layers do not typically have address information (there can be exceptions: an industrial network could have additional address information in the data field to be deciphered by the receiving station. This is an exception). These switches, however, can use other information to help determine network operation. This is typically called *policy-based networking* and can be used to prioritize data so that certain *types* of messages receive priority. For example, control messages can be given priority over data collection messages. Even so, layer 2 and layer 3 switches are the important ones for our needs.

Early routers were software based, relying on a general-purpose CPU (such as found in PCs and workstations) to run the software that handled the routing functions. However, traditional CPUs cannot keep up with the demands of a high-speed network. A router with even a few ports must process millions of frames per second. Today's routers and layer 3 switches use hardware-based functions, where the software and hardware coexist in special-purpose hardware. The most common form is an application-specific integrated circuit (ASIC), which is custom designed for the switch. Many ASICs, in fact, are custom designed by a company for their specific products. Such designs are no small tasks, involving thousands of man-hours. But the improvement in performance is dramatic, and ASIC-based devices are capable of keeping up with high-speed switched networks.

Another category of silicon device that has caught on in recent years is a network processor, which is a special-purpose processor that combines the benefits of an ASIC, but typically at a lower cost because it can be produced in high volumes for a variety of applications. While we think of ASICs as custom chips, some ASICs have become common enough that they are now off-the-shelf components. The difference between a custom ASIC and its off-the-shelf counterpart is that the ASIC often has manufacturer/device-specific software built into the hardware. The off-the-shelf network device either is programmable (allowing a manufacturer to customize it) or has generic software.

A layer 3 switch actually performs three functions:

1. *Path selection.* The switch determines the most efficient path from the sender to the receiver. Multiple paths often exist, and the switch uses a variety of tools to determine the best path, which can be the shortest or the most lightly loaded. One use of

this function is *load balancing* to prevent one path from being overloaded by heavy use while another path is underused.

The larger the network, the more paths that will be offered. For example, retrieving a Web page from a Internet site across the country will offer dozens of possible paths; routers used in the Internet backbone use sophisticated protocols to determine the best path.

On the factory floor, and in many companies, the number of paths is smaller and the task of the layer 3 switch is easier. The layer 3 switch is still a network device, not designed for heavy-duty Internet work. Still, even within a company, there will be multiple LANs that the switch will deal with.

2. *Forwarding packets.* This is the switching function of the router. It differs from layer 2 switching in that it can be forward based on both a network address and a station address within a network.

3. *Network services.* The switch can have a high degree of intelligence built into it so that it offers enhanced capabilities. These include network management, quality of service (making sure real-time applications like video get the proper priority), policy setting and enforcement, creation of virtual LANs, and so forth.

Switches are network specific. You buy an Ethernet switch. However, most switches will accommodate multiple higher-layer protocols so they can be used with the widest range of applications. For example, a layer 3 switch, depending on its complexity, can handle not only TCP/IP, but also protocols for Novell, IBM, AppleTalk, and Microsoft. It may even be upgradeable to handle future protocols. In addition, layer 3 switches can handle a variety of routing protocols that are outside the scope of this book. Nevertheless, you should realize that there are numerous protocols that *must* be supported, additional protocols that *should* be supported, and even lesser-used protocols that need to be supported only in niche applications. Wide protocol support is a marketing advantage, even if these protocols are never used. Many organizations may have 90 percent of its traffic using a popular protocol, but 10 percent may still be using that odd, outdated, or less-than-popular protocol. If you have an important application using a certain protocol, you do not want to reinvest in changing the application just to use a more common protocol.

The factory is a perfect example of the influence of protocols on future decisions and on the adoption of Ethernet. If a company has already invested heavily in ControlNet, decisions on additions or upgrades will be heavily weighted toward ControlNet. A competing protocol will have a tough—perhaps impossible—task of showing why a switch offers economic advantages. Even a so-called greenfield installation—one that is new from the ground up—will lean toward ControlNet, since the company has already invested large amounts of time and money in training for ControlNet. This is simply human nature: You lean toward what you already know unless you have a compelling reason to change.

This is one reason there are so many variations of industrial Ethernet. Manufacturers of automation equipment have a strong inclination to support existing protocols and hardware. Companies using the equipment have an equally strong inclination to continue using the protocols and hardware. The move to industrial Ethernet is partly driven by the desire of manufacturers and users to marry existing application protocols with TCP/IP and Ethernet. From one point of view, all use Ethernet and TCP/IP, so all run over common, standard Ethernet. Yet they use different upper layer protocols and applications which prevent them from talking to each other.

A narrower range of protocol support by a switch will lower the price of the device. In some ways, the market is divided between bare-bones devices that give you the basic necessary functionality and other devices with various degrees of bells and whistles to differentiate the device from its competition. We will be discussing some of these bells and whistles in later chapters.

Ethernet Devices: A Range of Stuff

The popularity of Ethernet means a wide range of Ethernet devices is available. Most of them are commercial equipment intended for the office environment. A growing range of industrial-grade equipment is available. Equally important, Ethernet ports are becoming standard equipment on many industrial devices, including PLCs and I/O units, making it easier to build a network without add-on fixes. Figure 5–8 shows an example of an I/O unit with built-in Ethernet.

Hubs form the least expensive approach, offering the lowest cost per port, although switches can be quite affordable. Industrial hubs and switches typically contain a low number of ports, with 4, 8, and 12 being common. They are DIN rail mountable and use +24 VDC power wired through terminal blocks, making them highly compatible with standard automation practices. Figure 5–9 shows DIN-rail-mounting industrial switches.

In commercial equipment, we often distinguish between workgroup and enterprise switches (Figures 5–10 and 5–11). The workgroup switch typically aims at lowering the cost

FIGURE 5–8   Ethernet-enabled I/O unit, which accepts a variety of plug-in modules
(Courtesy of Opto 22)

FIGURE 5–9   Industrial DIN-rail mounted switches. The one on the left has fiber ports in the upper right. (Courtesy of Phoenix Contact)

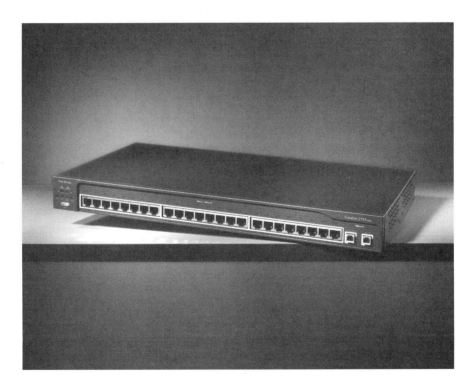

FIGURE 5–10   Workgroup switch
(Courtesy of Extreme Networks)

FIGURE 5–11 Enterprise switch
(Courtesy of Extreme Networks)

per port. It fulfills the functions of an Ethernet switch, but does not contain a lot of bells and whistles. Typical port counts are from 8 to 24.

Besides offering more ports, enterprise-class switches are more complex devices, offering more advanced services. For example, quality of service requires that the switch recognize different types of transmissions and give priority to certain types. To do so, the switch must have multiple memory queues to contain different types of priorities. How many queues and how much memory each contains all affect the complexity of the switch. In addition, most enterprise switches are chassis-based units in which boards plug into a backplane. Again, this represents more complexity. Even increasing the number of ports that the devices switch between increases complexity.

In addition, enterprise switches typically have more built-in reliability, such as two power supplies—with one operating and one on standby in case the first one fails.

INDUSTRIAL GRADE VERSUS COMMERCIAL GRADE: IS IT TOUGH ENOUGH?

Since Ethernet was originally intended for an office environment and most Ethernet equipment today is designed for such applications, is Ethernet hardy enough for the factory? Yes and no.

When we say *factory environment,* several visions are conjured up. It can be hot, damp, and dirty, with robotic arc welding equipment, fast moving conveyors, and so forth. Add in noise and vibration, and our image of the factory is a physically demanding place, especially for equipment designed to withstand the rigors of the office. Of course, not all factories are abysmal places. But even if the tough conditions are only localized in a few spaces, they still can stress network equipment. Automation equipment, after all, was designed with the factory environment in mind. Not so for the great mass of Ethernet equipment.

We can distinguish between commercial-grade equipment suited to the office and industrial-grade equipment ready for the factory floor. In general, the differences are how well the equipment stands up to

- Temperature extremes
- Vibration
- Dust and other contaminants
- Industrial solvents

Figure 5–12 outlines some differences between commercial and industrial grades for Ethernet equipment.

Temperature Range

Industrial network equipment typically operates over a wider temperature range to accommodate the tougher conditions found in many factories. Unlike an office building, which has better ventilation and temperature control, it is difficult to control temperatures on the plant floor, especially when surrounding equipment may operate at high temperatures.

Characteristic	Industrial	Commercial
Operating temperature range	0°C – 60°C	5°C – 40°C
Connector	Screw-Down Sub-D RJ-45 Sealed RJ-45 M12 circular Fiber	RJ-45 Fiber
Built-in redundancy	Important	Optional
Industrial enclosure	Yes	No
Electrical	24 VDC	110 VAC
DIN rail mounting	Yes	No
Ruggedized cable	Yes	No

FIGURE 5–12   Commercial versus industrial equipment requirements

Connectors

The RJ-45 connector that is the workhorse of the network industry does not have the ruggedness to withstand the factory environment. It was not designed to withstand severe vibration or rough handling. Nor does it provide the sealing required in many applications to prevent moisture and dust from crossing the interface. Industrial network equipment uses alternative connectors, such as a sealed RJ-45 or the M12 cylindrical connector. We will discuss these further in Chapter 7.

Redundancy

Network equipment is often built with multiple power supplies and other features to lessen the possibility of equipment failure. With dual supplies, one operates, while the other is on standby if the first one fails. Because a single failure can bring an entire assembly line to a costly stop, redundancy is important in factory applications. Commercial equipment also has redundant features to keep mission-critical networks running. Redundancy is typically found on larger enterprise switches, while the smaller work group devices have less redundancy. After all, it is not a critical failure if an office loses use of e-mail and Internet for an hour.

Industrial Enclosure

An industrial enclosure provides better protection against dust and moisture. This also raises the issue of ventilation: Commercial equipment often has a fan to provide cooling of the electronics. If dust is an issue, the industrial equipment must operate without internal ventilation—without fans or openings for air.

Electrical Input

A great deal of automation equipment runs on 24-VDC power. This power is run to the devices and terminated on terminal blocks. The 120-VAC power that runs commercial equipment may not be conveniently located. Designers of factory automation systems are familiar with 24-VDC power and find it convenient to have their equipment use it.

DIN Rail Mounting

PLCs and other automation controls often are mounted on DIN rails. The devices tend to be compact. Commercial equipment mounts in 19- or 23-inch racks. Again, the issue is compatibility with existing automation practices, the comfort of automation engineers with DIN rails and similar packaging, and a wish to avoid unnecessary challenges.

Ruggedized Cable

In an office building, cables are not subject to harsh environments—high temperatures, humidity, harsh solvents, and rough treatment through vibration or handling. Industrial cables have a stronger jacket that is more resistant to abrasion or solvents.

Fortunately, the choice is not an either/or decision. Most factory floors do not present a single environment: There are several. What is more, the plant itself means more than the factory floor. There are offices or other protected areas in which equipment does not have to meet industrial-grade specifications. Commercial equipment can also be installed in an enclosure that will protect it from many hazards. The point is that you can mix: industrial-grade hubs/switches near the action and commercial-grade equipment in other, more environmentally benign places.

SOME NOTES ON TOPOLOGY

This section looks at some typical topologies for industrial Ethernet installations. The emphasis is on star-wired switch-based networks, since these are most common today. As you will see, industrial Ethernet also adapts itself to a ring structure, which adds redundancy for reliability.

Linear Network

Figure 5-13 shows a straightforward linear installation. Switches are cascaded, one after another, using backbone ports on the switch. Automation equipment, be it PLCs, I/Os, PCs, or other devices, is connected to station ports. The drawback is that a failure of a single backbone cable segment will segment the network. Switches and attached equipment on either side of the break will not be able to communicate. Cable failure is the most common cause of failure.

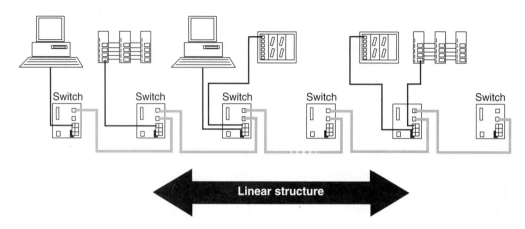

FIGURE 5–13   Linear network
(Courtesy of Phoenix Contact)

Ring Network

Ethernet does not recognize a ring network because it will not work under normal circumstances. When a signal is transmitted from a station in a shared-media network, it travels to all stations attached to a segment. In a linear network, there are endpoints clearly defined, beyond which the signal will not pass further. In a ring, the signal can continue around the ring, visiting stations a number of times and even traveling in both directions. A signal should pass over any cable segment only once. Otherwise an error occurs.

Still, in an industrial application, a ring structure adds reliability. Figure 5–14 shows a ring structure. Notice that the signal from one switch can travel in either direction to communicate with any other switch. If a cable breaks, the signal simply travels in the other direction to communicate.

While the figure shows a ring, it is not a true ring. One of the switches, sometimes called a ring manager, actually prevents the ring from being completed. It does not allow a signal coming in one side to exit in the other direction—unless there is a cable break or other malfunction in the path. In this case, the ring manager completes the path through itself, providing an alternative path for signals. You can think of the ring manager as being an electrical switch. Under normal conditions, this switch is open. If a failure is sensed in another part of the network segment, this switch is closed to complete the circuit.

The basic operation is like this. One side of the ring manager sends out a special signal to test the integrity of the network. The other side listens to receive the message. If it does receive it, it knows the network has no cable breaks in the backbone. If it does not receive the message, it assumes there is a cable break. It complete the path through the ring manager.

Figure 5–14 shows how the redundant paths work to provide greater reliability. Consider switches A and B, with switch C as the ring manager. Under normal circumstances, the signal from A to B must travel in a clockwise direction. It cannot pass through the ring manager. If a cable break occurs in the clockwise path between A and B, the ring manager will com-

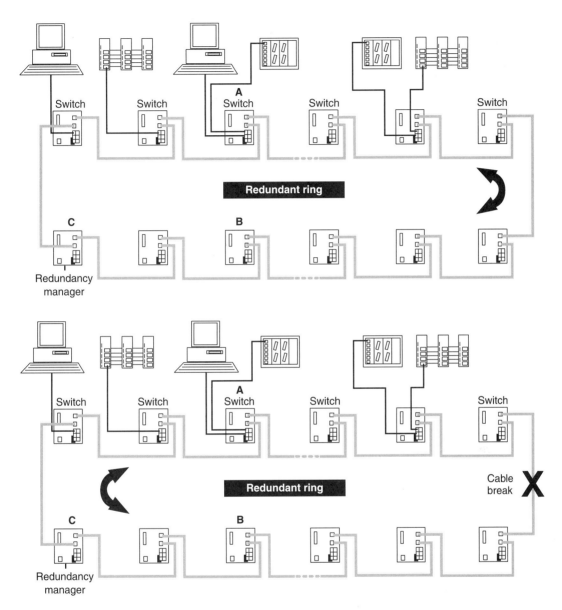

FIGURE 5–14   An Ethernet ring
(Courtesy of Phoenix Contact)

plete the path. Now the signals travel from A to B in a counterclockwise direction. Notice there are never two paths between switches A and B.

This ringlike structure, which is common with fieldbuses, is not part of standard Ethernet. But notice that it is not a real ring allowing signals to pass simultaneously in both directions to another station. It is really a linear structure that allows the path to be changed. What we have done is connect the two ends of a linear structure to the redundancy manager. A cable break actually splits the network into two separate networks. The task of the redundancy manager is to reconfigure the network into a single linear network again by providing a new path. Thus we are not really violating the Ethernet structure.

An important feature of the ring is how fast it can recover and reconfigure a network. In one real-world example, the redundancy manager can recover a 50-switch network in less than 500 ms. The time required depends on the capabilities of the redundancy manager, the physical size of the network, and the number of switches.

Figure 5–15 shows a more complex structure using multiple rings.

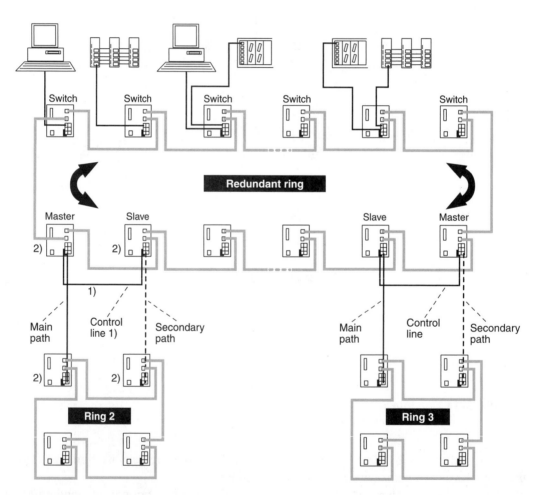

FIGURE 5–15   Multi-ring network
(Courtesy of Phoenix Contact)

SUMMARY

- The original Ethernet used a share-media topology.
- Switched, full-duplex Ethernet eliminates the determinism of collisions and offers a deterministic network.
- Principal Ethernet devices are repeaters, bridges, and routers.
- A layer 2 switch is a bridging device.
- A layer 3 switch is a routing device.
- Although most Ethernet topologies are star wired, some industrial Ethernet networks use a quasi-ring structure for reliability.

? REVIEW QUESTIONS

1. What is the speed of "slow" Ethernet? Fast Ethernet?
2. What are the three main types of cable that Ethernet uses?
3. What is the difference between a hub and a switch?
4. The original Ethernet ran over what type of cable?
5. Which is part of Ethernet proper: a MAC address or an IP address?
6. For a bridge and a router, which works at the MAC address level and which works at the IP address level?
7. What is the advantage of a ring network?
8. Does Ethernet use a true ring network?
9. Which of the following does not experience collisions:
   a. A shared media network
   b. A switched-media network using half-duplex transmission
   c. A switched-media network using full-duplex transmission
   d. Any network using full-duplex transmission
10. What is meant by redundancy in terms of a network?

Some Advanced Network Concepts: Managing the Network for Increased Performance

6

OBJECTIVES

After reading this chapter, you will be able to:

- List five uses of network management
- Define NMS
- Define the parts of an SNMP system
- Describe a virtual LAN
- Describe the role of a VPN in allowing remote access
- Describe the role of QoS in enhancing network operation
- Define how power is carried *over* Ethernet cables

MANAGED AND UNMANAGED NETWORKS

An Ethernet network can be managed and unmanaged. The simplest network is unmanaged. The stations, hubs, switches, bridges, and routers all perform their basic functions of delivering frames across the network to the intended destination. If properly planned and constructed, the network will work fine. After all, the network has enough intelligence to operate.

Several problems exist with an unmanaged network. You have no insight into what is going on. How much traffic? Who is sending the most traffic? Who is receiving the most traffic? What is the average frame size? Are there bottlenecks? Are there many collisions? Are there many errors, such as dropped packets? Do I need to upgrade the network to handle increased traffic? How do I know?

Network management brings monitoring and control to the network. Among the main capabilities of network management are these:

- Configure devices and the network
- Monitor network operation in real time
- Respond to problems and—more importantly—potential problems
- Analyze the network to plan future needs
- Manage security

Basic NMS with Simple Network Management Protocol (SNMP)

Network management, not surprisingly, is done through software. The most common **network management system (NMS)** is the **Simple Network Management Protocol (SNMP),** which defines a structure and set of rules for managing the devices on the network. A managed device is one that contains an SNMP agent. The agent has local knowledge of management information; it gathers the information about the device and organizes it in a form compatible with the NMS. You can think of the agent as an on-board computer that maintains a database for the device. Figure 6–1 shows the idea of a managed device.

The SNMP protocol defines a management database that describes the information that a device will collect. This includes both information about the device itself and information

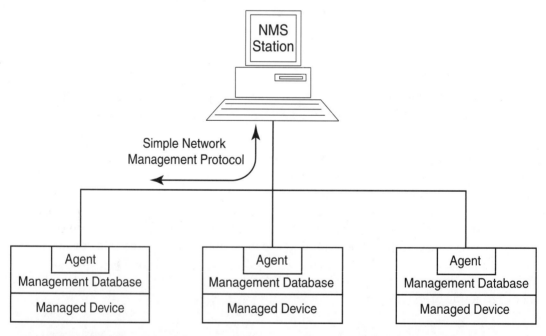

FIGURE 6–1 SNMP-managed device

about the traffic passing through the device. This information is defined in a MIB or management information base. The SNMP agent is the subsystem within the network device that monitors operation, manages the database, and communicates with the NMS software.

The Internet Engineering Task Force (IETF) has defined MIBs for many applications, such as Ethernet hubs, switches, and so forth, in documents known as requests for comments (RFCs). While managed devices will implement one or more of these RFCs, a company can also create a private MIB that contains additional or proprietary capabilities. For example, RFC 1156 defines MIB I, which was a standard MIB containing 114 objects. MIB II, defined in RFC 1158 and RFC 1213, supercedes MIB I and defines 185 objects. An object is an entry in the database. It can be a fixed value, such the device's name, or it can be a variable, such as the number of frames passing through the device in a given time. MIB II defines a management database for TCP/IP networks. That is, it concerns itself with packets.

SNMP offers four basic types of commands to monitor and control a managed device: *read, write, trap,* and *traversal* operations.

READ
The *read* command allows the NMS to get information from the device. It examines the values of variables in the agent's database. The command is used to monitor the managed device.

WRITE
The *write* command allows the NMS to change variables in the managed device's database. For example, a variable might control whether a given port is enabled to accept traffic or disabled. The write command allows the state of the port to be changed from enabled to disabled (or vice versa). Thus a write command allows the device to be controlled.

TRAP
The *trap* command is used by the device to report an event to the NMS. An event, in simplest terms, is a condition in the managed device that the NMS wants to be aware of as soon as it occurs. For example, a trap can be issued if the number of collisions exceeds a certain level, if a station transmits incorrectly formed frames, or if the number of bytes through a switch exceeds a certain level. In other words, the NMS is saying to the managed device: "if this condition exists, let me know." Traps are widely used to send alarms to the NMS.

TRAVERSAL
A *traversal* operation allows the NMS to determine which variable a managed device supports.

Monitoring can be done either at the MAC level, counting and measuring Ethernet frames, or at the network level, analyzing TCP/IP traffic. SNMP uses UDP.

Figure 6–2 lists typical information that the management agent will gather. This information will be supplied to the NMS station, which has the capability of analyzing it and displaying the analysis in various formats. An NMS uses a graphical user interface, such as Windows or a Unix graphical interface, to display information. This allows graphs to clearly

System type (hub, router, etc)
System name
Number of interfaces on device
Interface info (type, speed, address, etc.)
For each interface:
    Number of bytes received
    Number of unicast frames received
    Number of non-unicast frames received
    Number of good frames received and discarded (so as to free up buffer space)
    Number of bad frames received and discarded because of errors in packet
    Number of frames received and discarded because of an unrecognized protocol
    State of interface (on/off/testing)
    Number of bytes transmitted
    Number of frames transmitted
    Number of non-unicast frames transmitted
    Number of frames discarded before being transmitted
    Number of frames that could not be transmitted because of errors
    Number of collisions
    Number of excessive collisions
Status of components (such as fans, memory, etc.)
Temperature (alarm if too high)

FIGURE 6–2   Typical information gathered by the NMS

display network-operating statistics and trends. Because the NMS stores the information collected from the network, you can define the trends you want to view and the period for which
you want to view them. Note that Figure 6–2 expresses the basic digital unit as a *byte.* Most
RFC standards use the term *octet,* which is eight bits or a byte.

Industrial NMSs

Industrial Ethernet applications are greatly different from traditional architectures found in
office environments. Unlike the office in which all wiring is brought back to central switching closets, industrial applications have switches and hubs distributed in equipment enclosures and on the machinery itself. The distribution of equipment follows the earlier network
designs of the fieldbus technology. Network management software for industrial applications
provides the user with a true visualization of the actual connections and hierarchy. An example of this software is the package developed by Industrial Communication Technologies of
Newburyport, MA. This package is targeted at the industrial user who has a controls back-
ground and is responsible for management of industrial Ethernet networks.

    In addition to presenting a graphical representation of the connected devices, the software
interfaces with SNMP and accesses MIB data contained in managed devices (switches, hubs,
linking devices, and end nodes). Interconnecting lines change color (red-yellow-green) to
indicate status of communication at each link. The software can also display the connected
nodes as IP addresses, device names, process names, or locations. In Figure 6–3 you can see
a display with process names and that at Machine Center One, the PLC and I/O have red lines

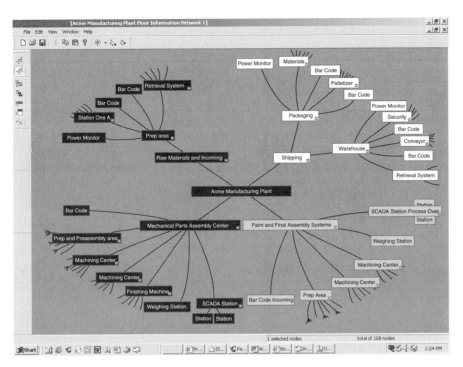

FIGURE 6–3   ICT global's visualization package showing network connections with process names
(Courtesy of Industrial Communication Technologies)

indicating that this system is down along with communication to the weighing station. This simplified network management software provides a basic level view of the network performance. Selecting a specific node will provide the ability to see IP address, MAC address, location, process, device type, as well as provide a means to FTP and Telnet to the selected node.

In addition to providing a graphical view of the industrial Ethernet network, the software allows for attaching pictures of either the device or machine at each level. These pictures provide a greater visual and provide plant floor technicians with the ability to quickly identify the equipment or device in question. As seen in the following picture, the software can also directly link to Web pages embedded or linked to each device. Embedded Web pages are becoming a standard offering by many of the automation equipment suppliers. Web pages are brought up as windows on the main screen.

The network management software is run continuously, and network irregularities are logged on a 24×7 basis. The content of the linked Web pages can include such information as machine performance, process performance, user manual, and wiring diagrams. Information can be viewed through a simple browser and provide an organized method to manage and support the network and connected devices.

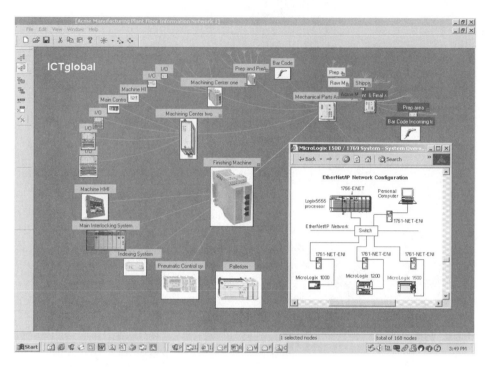

FIGURE 6–4   Sample detailed display with Web page opened with local configuration at a specific machine
(Courtesy of Industrial Communication Technologies)

Benefits of Management

Network management has several advantages in allowing you to control and monitor a network. These include analysis, alarms, and control.

ANALYSIS

By building a history of network operation, you can achieve several levels of efficiency. First is trend analysis. You can see if traffic is growing or declining, if errors are becoming more prevalent, and if there are bottlenecks that are bogging down the network. In an office, for example, the network might experience a significant increase in traffic and attendant decrease in performance during the lunch hour, when employees use free time to surf the Internet. In one factory application, improper operation of the network was traced to the time of day when servers were backed up. The backup consumed so much bandwidth that the rest of the network could not keep the automation equipment functioning properly.

With the information gathered from trend analysis, you will be able to fine tune network topology. Too much traffic here? Consider adding a bridge to segment the network. Too many collisions there? Maybe you need to replace hubs with switches. A slowdown over there?

Upgrading from 10 Mb/s to Fast Ethernet might solve the problem. What effect is a new routine in the PLC having on the network operation? Without network management, you will have very little insight into the operation of the network.

Analysis allows better capacity planning. Trends will help you anticipate growing needs. Do the trends suggest you need to expand or upgrade the network soon? In six months? In a year? Such forecasting also eases the budgeting for capital expenditures and ensures factory uptime and a clearly defined path for growth.

ALARMS

An important feature of an NMS is the ability to provide alarms to prevent network failure and costly downtime. You can typically specify two levels of alarms: notification and alarm. For any given MIB object that you can set an alarm for, you can usually set two thresholds for which you want the NMS to inform you. One level is notification: The SNMP agent simply sends you notice that a threshold has been passed. The second level is an actual alarm, typically with both audible and visual notification. Think of the two levels as (1) it is about to break and (2) it is breaking.

The ability to set multiple thresholds for alarms is a key to proactive management. You can anticipate problems, look into them, and fix them before they disrupt operation. One example is number of bad frames. Ever so often, a bad frame may be received. While Ethernet is very good at transmitting good frames, an increase in bad frames could indicate a problem—anything from overheating in a switch or station to cable faults. You can set a threshold of eight bad frames for notification and twelve bad frames for an alarm. When either threshold is reached, you will receive either a notification or an alarm.

Some devices have built-in features to monitor temperature and the health of basic components such as fans and power supplies. Again, the ability to be notified when a component is heading south or to receive an alarm is a great boon to reliable network operation.

This monitoring and alarm function often also allows you to set rules to recover from a problem. Do you want the system to shut down? Do you want it to attempt to repair itself? Do you want to disable a port that has a malfunctioning device attached to it? Can you reroute the data?

The more robust management system not only allows alarms to be displayed (visually and audibly) on the NMS station, it also permits them to be sent just about anywhere you specify. It can use dial-up networking to phone a remote location. It can call your cell phone or beeper. It can e-mail you. The point is that you never need to be out of touch with the network, caught by surprise.

CONTROL

The NMS will also allow you to control the network. For example, you can enable and disable ports on a switch. This will allow you to lock users out of the network if they cause problems. Maybe you want only this PLC or that PLC to be active on the network at a given time. The NMS allows you to enable or disable the ports to which each are attached.

VERSIONS OF SNMP

There have been three versions of SNMP, conveniently called SNMP, SNMPv2, and SNMPv3. Each builds upon the earlier versions.

SNMPv1

SNMP, described in RFC 1157, is the original version of SNMP and the de facto management protocol for Ethernet and the Internet.

SNMPv2

SNMPv2 is an evolutionary advance of SNMP. Basically, it beefs up the communication between the agent and the NMS. For example, a new capability statement indicates the level of support the agent gives a MIB group. The NMS will adjust its behavior according to the capability statement. SNMP adds two operations: *GetBulk,* which allows it to retrieve large blocks of information, and *Inform,* which allows one NMS to send a trap to another NMS.

SNMPv1 and SNMPv2 are not compatible: They use a different header and data (called the protocol data unit or PDU). Two ways to maintain compatibility between the two versions are proxy agents and bilingual NMSs. In the proxy system, the agent translates SNMPv2 commands into SNMPv1 commands. When reading SNMPv1 data, it translates these into SNMPv2-compatible data. A bilingual NMS simply determines whether an attached device is using SNMPv1 or SNMPv2 and then communicates using that protocol. In other words, it communicates with a managed device in its "native" protocol.

SNMPv3

SNMPv3 adds a level of security, authenticating that a transmission originates with the authorized device and encrypting the transmission.

One drawback to SNMP is that it works by polling each device at regular user-defined intervals. This polling uses up network resources: The bandwidth used by SNMP polling is not available to carry traffic from the automation system. On the one hand, you want to poll as often as possible to keep up-to-date information as fresh as possible and to get alarms as soon as they occur. On the other hand, you want to poll only occasionally so as not to clog up the network with management traffic. Particularly in parts of the network that must operate deterministically, management traffic must be factored into the network's operation and the polling interval judiciously picked.

RMON is an alternative to SNMP that helps overcome this polling limitation.

RMON

RMON stands for remote monitoring. What it does is transfer the monitoring responsibility to the managed device, rather than the NMS station. The device collects the data on traffic and monitors for alarm conditions. In the absence of an alarm, RMON sends collected data to the NMS station only occasionally—perhaps only once an hour. This eliminates the need for polling by the NMS. And because the RMON agent in the device does some of the rudimentary analysis, the information provided is more meaningful and requires less processing by the NMS station. It also saves bandwidth.

What about alarms? If a threshold is reached, the RMON will immediately alert the NMS station of the condition. It does not have to wait for a regularly scheduled transmission. This can mean an even faster response than SNMP polling.

While they are quite powerful and useful, SNMP and RMON form the most basic network monitoring and control. Their basic function is to allow you to monitor traffic and proactively identify and correct potential problems. Every network, beyond the most basic, will benefit from network management—and it is additionally useful in factory automation where real-time operation is critical. You do not want changes in network operation to go undetected until operation is no longer in real time. A basic NMS, using SNMP or RMON, makes no distinction between types of traffic on the network. An Ethernet frame is an Ethernet frame. The NMS can tell you how big it is, whence it comes and where it is destined to go, if it is correctly constructed, and so forth. But it can not tell you whether it is SNMP data, publish/subscriber data from the factory network, an upload of production data to the scheduling department, and so forth.

RMON defines eight groups of functions:

- *Statistics:* counters for packets, bytes, broadcasts, and errors
- *History:* periodic samples of statistics group counters for later transmission
- *Hosts:* statistics on each host device on a segment or port
- *Host Top N:* a user-defined subset of the hosts group, used to keep management traffic minimized
- *Traffic matrix:* statistics of conversations between hosts on a network
- *Alarms:* user-defined thresholds on critical variables
- *Events:* SNMP traps and log entries when an alarm group threshold is exceeded
- *Packet Capture:* manages packet buffers

Many RMON-equipped devices do not implement all eight groups in order to reduce the cost and complexity. Typically statistics, history, alarms, and events are the groups implemented.

RMON 2 adds several more groups:

- *Protocol directory:* protocols that the SNMP agent monitors and maintains statistics about
- *Protocol:* statistics on each protocol
- *Network layer host:* traffic statistics for each network layer address on the segment or port; in this case the IP address
- *Network layer matrix:* traffic statistics for pairs of network layer addresses—who is talking to whom
- *Application layer host:* statistics by application layer protocol for each network address
- *Application layer matrix:* traffic statistics by application layer protocol for pairs of network layer devices

- *User-defined history:* a way of extending the RMON 1 history group to include link layer statistics to include any RMON, RMON 2, MIB I, or MIB II statistics
- *Address mapping:* MAC to network layer address binding
- *Configuration:* SNMP agent capabilities and configurations

In factory-floor applications, the rich set of statistics offered by RMON 1 and RMON 2 may be overkill and may present transmission overhead that runs counter to the determinism of the automation network. Careful consideration of the types and frequency of information required can help make management useful.

There are, however, additional management capabilities offered by modern Ethernet. These include policy-based management and quality of service. These require greater understanding of the data inside the Ethernet frame.

SMON

SMON—for switch monitor—adds specific control and monitoring features for switches.

VIRTUAL LANS

There are several reasons to segment a LAN, to break a larger LAN into several smaller ones. Most reasons involve traffic management in an effort to make traffic over the LAN as efficient as possible. Suppose, for example, you have several development stations on your factory LAN. The LAN itself may have several segments, each handling a different manufacturing cell or operation. You may want to have the development station attached to this segment today and another segment tomorrow.

Physical Segmentation

The traditional method of segmenting a LAN is by segmenting the hardware. You can add a bridge. Or you can physically re-arrange connections; you can remove the station from the switch on this side of the LAN and plug it into the switch on the other side. In other words, you segment your LAN physically.

Logical Segmentation

A **virtual LAN,** or **VLAN,** allows you to create logical segments rather than physical ones. A simple example is a VLAN switch. It allows you, for example, to specify that ports 1, 2, 3, 7, and 9 belong to LAN A. Ports 4, 5, 6, and 8 belong to LAN B. And ports 10 through 24

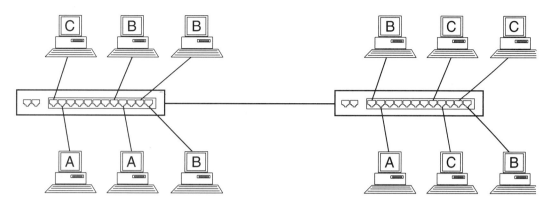

FIGURE 6–5   A VLAN allows stations to be assigned to a LAN segment by address, port, or protocol. Here each user is part of one of three LANs: LAN A, LAN B, or LAN C.

belong to LAN C. The real value of the VLAN comes into play when you want to rearrange the LAN. Through software, you can easily reassign each port to a different LAN. Figure 6–5 shows the idea of a VLAN. In this case, users are assigned to one of three LANs. Notice that the LAN assignment is not based on physical location. Equally important, the assignments can be easily changed whenever a change is warranted.

You can define a group of switch ports into a community of interest. One reason is to form a team of workers devoted to a project. When the project is complete, the VLAN can be re-arranged for the next project. Maybe a team is working to program or debug a manufacturing line. They are physically connected to both the development station and to the factory equipment. Once the line is function properly, the team can be connected to another line, for example.

A VLAN can also be used to separate a high-security application from a lower security one.

To support a VLAN, the Ethernet frame has an additional VLAN tag added.

The VLAN creates a single broadcast domain in switched-media networks. The VLAN can be extended beyond a single device. Also, a VLAN can be created over a wider geographical location. The developers and the PLCs do not have to be connected to the same switch or even be in the same building.

You can assign a VLAN based on different criteria:

- *Port driven.* Users are assigned by the ports to which they are connected. This is a simple, straightforward approach that is conceptually easy to visualize. The VLAN assignments are easily viewed on a graphical display. Each port is assigned to only one VLAN at a time.
- *MAC address driven.* The VLAN assignments are based on the MAC (Ethernet) address of the attached station. This is somewhat more difficult, since a table must be created for each VLAN. If the VLAN is large enough, the table will have to reside on

a server rather than in the switch. But the approach does offer greater flexibility. For example, a laptop user might end up connecting through different switch ports depending on his location.

- *Network address driven.* The IP address can also be used with similar advantages and disadvantages.
- *Protocol driven.* Assignments can be based on the type of protocols carried (at any level of the OSI model).

VIRTUAL PRIVATE NETWORK

First off, do not confuse a **virtual private network (VPN)** with a virtual LAN. A VLAN is a method of logically grouping users. A VPN is a method of using a public network as if it were a private network. The most common example today is using the Internet to extend a company's private network to remote users and remote locations. In short, you are using the public Internet as if it were part of your private network. You can extend your network to remote offices, mobile users, telecommuters, business partners, and anyone else you wish. The main goal is to achieve the same level of security as you enjoy in a private network not attached to the outside world.

Key Concepts

A VPN relies on four key concepts:

- Tunneling
- Encryption
- Security
- Quality of Service

For a VPN to be successful, it must be secure from prying eyes. Malicious hackers and even the mildly curious can intercept a packet and learn confidential information that may be inside. Besides sensitive information, such as passwords and user IDs, the hacker can learn addresses of servers. Armed with the right information, a hacker can pose as an authorized user and run amok in your network. He can steal additional information or disrupt the factory. This is one reason many network administrators fear the Internet (especially since they cannot always control careless users).

The VPN offers a secure method of encrypting transmissions so that they can be sent in the clear over the Internet. The VPN does this by encrypting your packet, including any IP header information containing source and destination addresses. It then encapsulates this information into another packet with an unencrypted header to allow the packet to be routed. See Figure 6–6.

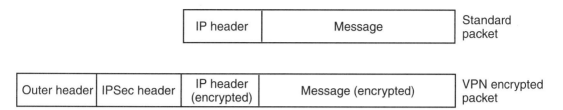

FIGURE 6–6 VPN encapsulation

TUNNELING

When an IP packet is thus encrypted and encapsulated, it is said to be in the *tunnel* mode. Conceptually, you have built a tunnel in the public Internet infrastructure. As the packet travels across the Internet, the "real" information is hidden from the routers and other devices that manage traffic. It is also hidden from hackers. A hacker intercepting the packet will find only gibberish. And the IP address yields nothing useful, since it is not the ultimate destination or source. At the destination, a VPN gateway accepts the frame, removes headers and recovers the encapsulated data, unencrypts it, and processes it in a normal manner.

ENCRYPTION

Encryption is a complex subject, but we need to touch on a few points. It used to be that encryption standards were kept secret. The drawback is that once a hacker cracked the code, he had access to all the information. How good is the encryption technique? Sometimes, it is hard to tell because a secret technique is not subject to rigorous public testing.

It is more common today for encryption algorithms to be published so that everybody knows how they work. The difference is that there must be a key—a series of bits that can unlock the encrypted message. The longer the length of the key, the more secure the encryption. Think of a combination lock. You can, conceivably, try every possible combination until you finally unlock the lock. But the sheer number of combinations makes that a near-impossible task. An 8-bit security key has only 256 combinations. Even a slow computer can work through all those combination in short order. A 16-bit key has 65,536 combinations—more complex, but solvable in a short time with a computer. Some VPN products use a 168-bit key, which provides 3.74×10^{50} combinations. Even supercomputers would need years to get though all the possibilities. And remember, even if the key is cracked, it can be changed to a new combination. The point is that VPNs are very secure.

The most common encryption algorithm is the Data Encryption Standard, or DES. It uses a 56-bit key to encrypt data into 64-bit blocks. A 56-bit key allows 72,057,594,037,927,900 combinations. It would take about 20 years for a personal computer to run through all possible combinations. Thus, the encrypted data are safe from all but the most determined and well-funded snoopers.

The 168-bit key mentioned earlier is achieved by taking your data, encrypting it with a 56-bit key, decrypting it with a second key, and then re-encrypting it with yet a third key. This method is known as 3DES.

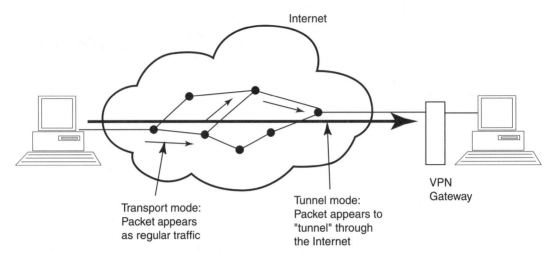

FIGURE 6–7 IPSec modes

SECURITY: VPN PROTOCOLS

Three protocols are used for VPN applications: IPSec (IP Security), L2TP (Layer 2 Tunneling Protocol), and PPTP (Point-to-Point Tunneling Protocol).

IPSec is the security standard for the Internet. It defines two modes (Figure 6–7): transport mode and tunnel mode. The transport mode does not encrypt the source and destination address; the tunnel mode does encrypt them. The transport mode is suited to use over a LAN, while the tunnel mode is more appropriate for the Internet.

L2TP is useful for transmitting non-TCP/IP protocols, such as NetWare IPX or AppleTalk. Still, L2TP uses IPSec for encryption.

PPTP is a proprietary encryption and authentication protocol devised by Microsoft. While PPTP was to have been replaced by L2TP, it is still used by Microsoft to support VPNs in Microsoft Windows. Unlike IPSec, PPTP does not use DES for encryption. Instead it uses RSA (named after the inventors, Rivest, Shamir, and Adelman), which is not as robust as 3DES.

VPNs make a convenient method for remote monitoring of sites.

QUALITY OF SERVICE

We mentioned Quality of Service (QoS) as one requirement of a VPN. But QoS is not simply related to VPNs; instead, it is a characteristic that is growing in importance in Ethernet networks.

The basic idea of QoS is to provide guaranteed delivery of certain types of data. You must be able to prioritize transmissions and give priority to certain types of transmissions. Many types of data are crossing the network at any given time. Some are more important than others. For example, commands to actuators and input from sensors must be delivered in a timely manner. Historical data uploaded to the enterprise are not nearly as time critical and do not require the same priority. In the same way, screens of an HMI downloaded from a server are not as critical as real-time displays of operations. Large file transfers, which can really hog bandwidth, can be given a low priority.

But QoS goes beyond the determinism we have so far discussed with Ethernet. From the factory viewpoint, as important as determinism is, it is the lowest form of QoS. More important is to give priority to certain types of transmissions.

QoS is important for supporting real-time operations, whether it is voice, video, or the operation of factory equipment. QoS is required wherever there is the potential for congestion and the delay of time-sensitive information.

Key Transmission Characteristics

There are three basic transmission characteristics that are typically key to QoS:

LATENCY

Latency is the time between a node transmitting a message and another node receiving it. For a device such as a switch or router, it is the time between a message being received on the input port and being transmitted on the output port; it is the time it takes the device to process the message and pass it along.

JITTER

Jitter is the variation of a signal in relation to its original reference time. Packets arrive at different intervals than those that they were transmitted. Jitter causes a jerky, shaky quality in video.

BANDWIDTH

Bandwidth. If sufficient bandwidth is not available, timely delivery of messages cannot occur. Even in an efficient full-duplex switched LAN, too much traffic can cause delays, allow devices to drop packets, and require retransmission of information.

LOST PACKETS

Lost packets occur when a device cannot keep up with traffic. If a switch cannot process packets as fast as they arrive on a port, packets will be lost.

For example, a multimedia application requires that video be delivered at a steady pace and with little variation in time between delivery of frames. If the delivery is not steady, the video will be jerky: The more variance, the more jerkiness. Notice that the deterioration of video can occur for two reasons: insufficient bandwidth and high variance. Suppose a video requirement—say for training on setting up a machine—is 1 Mb/s. This is a modest requirement, especially in a switched fast Ethernet environment. But the bandwidth is required at all times when the video is being transmitted. We do not mean an average over time: This leads to variance.

QoS and Industrial Ethernet

Most QoS efforts center on Internet and similar applications. How do you send information over long distances and over various routed paths with the required bandwidth and low latency? While such needs occur in factory automation, they are not our main concern. In industrial Ethernet, the most fundamental QoS issue is to ensure that the automation equipment operates properly. This means that the network must accommodate the response intervals and scan times of the application.

Most QoS efforts fall into one of two categories: best effort and guaranteed.

BEST EFFORT

Best effort. The network uses a built-in mechanism to ensure the timely delivery of packets, but makes no guarantees.

GUARANTEED

Guaranteed. The network guarantees certain levels of bandwidth and latency to the application.

You saw the lowest level of QoS in discussing the evolution of Ethernet. By moving from a half-duplex, shared-media environment to a full-duplex, switched media environment, Ethernet eliminated collisions and gained determinism. While most experts would not consider this to be true QoS, from the needs of automation equipment it represents a giant step in quality of service.

Prioritization

The next step is prioritization of frames. Here certain types of frames are given a higher priority and passed through a switch faster than lower priority frames. Prioritization can be implicit or explicit. With implicit QoS, the network administrator can allocate service levels to each switch or to the entire network. This prioritization of service can be based on different criteria: type of application, protocol, source address, destination address, switch ports, and so forth. The switch will examine each frame to determine its priority.

One aspect of QoS is the increase in complexity of QoS-enabled devices. For example, they must provide queues so that transmissions of lower priority can be temporarily stored while higher priority messages are moved ahead. A network device, such as a switch, can have multiple queues for different types of data. Suppose a given port handles both time-critical sensor data and SNMP management data. The sensor data are put into a high-priority queue and passed through the port faster than the lower priority SNMP data, which use a low-priority queue. Figure 6–8 shows a simple example of a queue. Type A transmissions require real-time transmission, while Type C uses low-priority data requiring no priority. On the input side, the mix of incoming data types is relatively random. On the output side, you can see the prioritization: Every other frame out is type A.

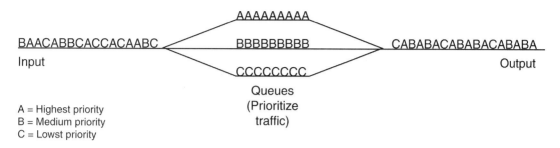

FIGURE 6–8   Priority queues help QoS

Queuing

Queues have a couple of drawbacks. Each queue must be large enough to store the frames while higher priority data are processed. This requires additional memory—enough so that the queue does not overfill and lose information. In addition, high-priority data must not be allowed to "capture" the port to its exclusive use so that low-priority traffic is never given a chance. Queues require that each frame be inspected to determine its type. This places additional processing burden on the device. (You could also prioritize by input port rather than data type. All transmissions coming in on port 1 are given a high priority, while all transmissions on port 5 are assigned a low priority. Similarly, prioritization can be based on output ports.)

In contrast to implicit prioritization, which is built into the network, explicit prioritization allows the user or application to request a certain level of service. The switches will make a best effort to provide the service level.

Version 4 of the Internet Protocol contains a field for IP Type of Service (TOS), which allows bandwidth, delay, and reliability attributes to be set.

RSVP

The Resource Reservation Protocol (RSVP) is an industry-standard protocol for reserving bandwidth along an end-to-end path. RSVP works on a per-application basis. In the video example, the transmitting station might transmit a reservation for 1-Mb/s bandwidth with a 200-ms delay. Each device along the path must agree to this reservation and notify the sending station that it can meet the requirement. Once end-to-end agreement is reached, the transmitting station begins sending the video. The potential drawback here is that all devices must be able to handle RSVP. Such approaches to QoS require end-to-end compliance in order to work. Otherwise, a non-RSVP device in the path becomes the weak link that ruins the whole chain of communication.

Notice that most approaches to QoS operate with TCP/IP at the network level. QoS can also be obtained within an Ethernet frame. The IEE 802.1p/Q standards define a method for inserting prioritization/QoS information into an Ethernet frame. 802.1, remember, defines the logical link layer (LLC), which is part of the data-link layer shared with Ethernet's MAC. In 802.1p/Q, three bits are allocated to specifying the priority of a frame. These three bits are part of the VLAN tag, which you will recall is a two-byte addition to the Ethernet frame. This gives eight possibilities, numbered 0 (lowest priority) through 7 (highest). If the switch does not have enough queues to support 8 different priority levels, it will group them, say 0-3, 4-5, 6-7.

Guidelines

Here some rough guidelines on how to achieve QoS in a factory network.

- Use switches and full-duplex transmission, rather than hubs, to eliminate collisions and increase determinism.
- Use low-latency switches—those that operate at wire speed or near wire speed so that traffic is not delayed in the switch.
- Use the highest transmission speed that makes technical and economic sense. Move from 10 Mb/s to 100 Mb/s for stations-to-switch connections and from 100 Mb/s to 1 Gb/s for switch-to-switch and server-to-switch links.
- Keep network segments small enough to be lightly loaded—50 percent or less.
- Use QoS-enabled devices to prioritize traffic or define service levels. Implicit prioritization is easier to use and more easily implemented, although explicit prioritization may be more flexible. Prioritization should be based on giving real-time automation traffic priority. For layer 2 switches, use 802.1p/Q prioritization. For layer 3 switches, use RSVP or another higher-level QoS.

POWERING REMOTE DEVICES: POWER OVER UTP

Traditionally, Ethernet does not allow power to be supplied over the copper UTP cable (described in the next chapter) that carries data. Office devices are powered by plugging them into 120-Vac outlets—either directly or through an external power adapter. Some industrial equipment uses this same approach, while other equipment can operate off the 24-VDC supplies common in factory applications. In this second case, field wiring to terminal blocks is most common. For Ethernet switches and other common devices mounted on DIN rails, this approach is sensible.

For remote devices, such as sensors or actuators, the need to provide additional wiring for power is less satisfactory. While traditional fieldbuses accommodate power over the same cable that carries data, Ethernet traditionally does not.

Under development as this book is being written is IEEE 802.3af, which provides a method to provide power over Ethernet UTP cable to remote devices. While some suppliers already provide power capabilities, these are nonstandard and do not comply with IEEE

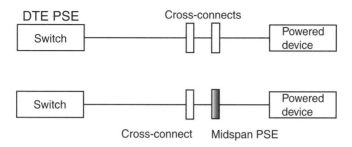

FIGURE 6–9   Power over Ethernet cable: DTE and midspan

802.3 requirements. The approaches are proprietary and lock you into a single-vendor solution. A standard method will open up a whole new class of low-power remote Ethernet devices in automation applications.

The standard allows two types of so-called power-sourcing equipment (PSE): Data terminal equipment (DTE) and midspan equipment (Figure 6–9).

A DTE PSE is a device such as an Ethernet switch. The DTE provides power on the same cable powers that carry data (pairs 1-2 and 3-6). It can be used with 10BASE-T, 100BASE-T and 1000BASE-T. The PSE provides 44 V to 57 V and a maximum current of 350 mA,

In midspan power, the power is inserted somewhere along the cable segment onto unused pairs (pairs 4-5 and 6-8). Midspan insertion cannot be used with 1000BASE-T. Midspan insertion provides two benefits: (1) it means that each port of an Ethernet switch does not have to provide power and (2) it allows use of legacy equipment that does not support power sourcing. Even so, DTE PSE applications are simpler to implement and more straightforward to support.

It is important that the power sourcing not degrade the performance of the cable segment. This becomes especially important with midspan insertion of power. Just as an interconnection such as a cross-connect or patch panel represents a potential point of degradation—a performance bump, as it were—so the midspan insertion represents yet another. The goal of the standard is to have the insertion be equivalent to a cross-connect. Many installations will replace a cross-connector or a drop cable so that performance is maintained. For example, commercial building standards for cabling put limitations on the number of outlets and cross-connects in a path; to maintain the limit, an application may be required to eliminate a cross-connect to allow midspan PSE to be used.

Power over Ethernet is advantageous to decrease the complexity of cabling to remote devices. In industrial applications, where remote sensors or actuators are inconveniently located, this reduction in wiring increases reliability and simplifies installation, maintenance, and troubleshooting. Because the cable segments will be short in terms of Ethernet cabling (which allows a maximum span of 90 meters for UTP) and often have fewer or no cross-connects, design for power is equally simplified.

Power, of course, cannot be supplied over optical fibers. In most cases, this will not be detrimental. Fiber is favored for longer spans, higher data rates, or where electrical cable is hazardous; it is not widely used between a switch and remote device and so does not "compete" with copper UTP in such applications.

SUMMARY

- A network management system allows an Ethernet to be monitored and controlled.
- The most popular management protocol is SNMP—Simple Network Management Protocol.
- A virtual LAN allows LAN segments to be formed and re-formed through software.
- A virtual private network provides a secure way to achieve remote access to a network.
- Ethernet devices can be powered over UTP cabling, which can simplify cabling in industrial applications.

? REVIEW QUESTIONS

1. What are the two basic functions of an NMS?
2. What management protocol is commonly used with an NMS?
3. How does polling by an SNMP agent potentially degrade network operation?
4. How does RMON overcome the limitations of polled SNMP management?
5. What is the main value of a virtual private network in allowing remote access?
6. What is the standard security protocol used for VPN access over the Internet?
7. Define quality of service.
8. How does queuing help achieve QoS?
9. What are the two methods of providing power over Ethernet cable called?
10. On which type of cable can power be supplied?

Cables and Connectors

7

OBJECTIVES

After reading this chapter, you will be able to:

- Distinguish among the three main types of copper cable: coax, UTP, and STP
- Define the major difference among categories of UTP
- Distinguish between the two major types of optical fiber
- List the main types of connectors used for copper and fiber Ethernet applications
- Give basic distance rules for cable runs

CABLES: THE ETHERNET MEDIA

Cables, naturally enough, are vital since they are the medium that connects all the various equipment together and provides the path over which signals travel. There are wireless versions of Ethernet, but we will cover them separately in Chapter 8.

As we mentioned earlier, cable technology played an important role in the evolution of Ethernet. As Ethernet increased in speed and moved from cumbersome coaxial cable to low-cost twisted-pair cable, it was essential to ensure that twisted-pair cable had sufficient bandwidth to accommodate higher data rates. The evolution of Ethernet goes hand-in-hand with the evolution of cables. In addition, fiber-optic cable offered even greater bandwidth and transmission distances, although high costs and unfamiliarity kept it from dominating cabling systems.

COPPER CABLE CHARACTERISTICS

In this section, we will be discussing the properties of copper cable that affect signal transmission and the ability to transmit a signal without undue impairment. In terms of importance, the most fundamental are:

- Attenuation
- Characteristic impedance
- Near-end crosstalk (NEXT)

There are numerous additional characteristics that distinguish cables from one another, but we will concentrate on attenuation, characteristic impedance, and crosstalk. Understanding these will clarify many of the differences among copper cables.

Attenuation

Attenuation is the loss of signal power along the length of the cable. A significant feature of attenuation in copper cable is that it increases with frequency. Attenuation per unit length is greater at 10 MHz than at 1 MHz. Figure 7–1 shows typical attenuation for coaxial and twisted-pair cables.

Attenuation must not become severe enough that the receiver can no longer distinguish the signal from the noise. Attenuation significantly limits practical cable lengths, especially at high speeds where attenuation is more severe.

Noise

When the first personal computers were introduced in the late 1970s, they radiated enough energy to interfere with televisions several hundred feet away—not just in the house where they were used, but in neighboring houses as well. This interference is known as electromagnetic interference (EMI). EMI is a form of environmental pollution with consequences ranging from the merely irksome to the deadly serious. EMI making television reception snowy is one thing; its interfering with a heart pacemaker is quite another.

A copper wire can be a main source of EMI. Any wire can act as an antenna, radiating energy into the air where it can be received by another antenna. A simple demonstration of EMI can be done by placing an AM radio near a computer. You can tune into the computer at some point on the dial. Try running different programs to hear the differences each makes. What you are hearing is EMI.

Just as a wire can be a transmitting antenna, it can be a receiving antenna, picking up noise out of the air. Obviously, this received noise can interfere with the signal being carried. Of special significance is noise that couples from one wire to another in the same cable.

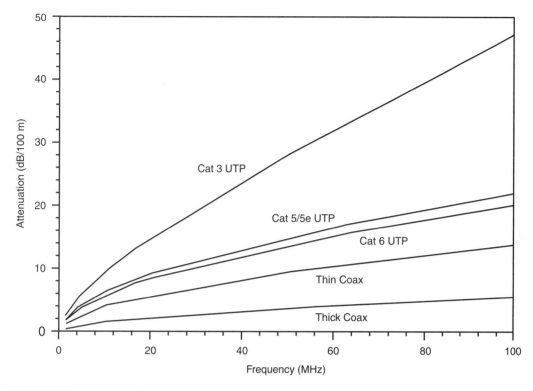

FIGURE 7–1 Attenuation for coaxial and UTP cables

Factories can be electrically noisy environments, and some factory equipment presents special problems with electrical noise. Arc welders, variable-frequency drives, and motor starters are examples of equipment that can emit high-frequency electrical energy. Properly designed and deployed cables offer excellent noise immunity. Still, the electrical environment of the plant floor may mean industrial-grade cables can be used in place of commercial-grade office cables. In particular, the category 5 and 6 twisted-pair cables used in the office are not the best choice. As we will discuss, a new industrial version of category 5 cable (called Cat 5i) has been developed that is more compatible with the physical and electrical environment of the factory.

Crosstalk

A special type of noise is crosstalk. Crosstalk is energy coupled from one conductor to another in the same cable or between cables. Crosstalk was once fairly prevalent in telephone systems, where you could faintly hear another conversation.

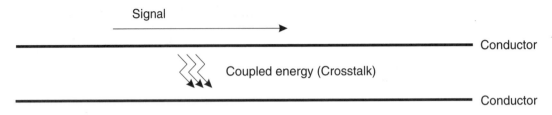

FIGURE 7–2 Crosstalk

When current flows though a wire, it creates an electromagnetic field around the wire. This field can induce a current in an adjacent wire within the field. This coupling is crosstalk.

Figure 7–2 shows the idea of crosstalk. A signal is transmitted down a wire (called the *active* or *driven line*). A certain portion of the energy couples on a second wire (called the *quiet line*). This energy on the quiet line is the crosstalk. Crosstalk is measured either in percentages or decibels. In percentages, the magnitude of the signal on the quiet line is given as a percentage of the signal on the driven line. Thus, the crosstalk might be 3% or 10%. If the driven line signal was 5 volts, a 10% crosstalk means that 0.5 V appeared on the quiet line. In communication cabling, crosstalk is more often expressed in decibels, a term we will turn to in detail shortly.

Crosstalk can be measured at the near end or far end of the line. In near-end crosstalk (NEXT), the energy on the quiet line is measured at the same end as the source of the signal. In far-end crosstalk (FEXT), the energy is measured at the opposite end, away from the signal source.

NEXT

NEXT is of interest in every application of twisted-pair cable. Consider a signal on the driven line. As it travels down the wire, it becomes attenuated, losing energy. At the opposite (far) end, it is weakest. So the greatest amount of energy is available for coupling at the near end. Similarly, crosstalk on the quiet line is usually strongest at the near end for two reasons. First, it can receive the greatest energy here since energy on the active line has not yet become attenuated and, second, it will not be attenuated as it travels to the far end.

NEXT has an additional practical consequence. In a typical network, one pair of wires is used to transmit, while another pair is used to receive, as shown in Figure 7–3a. The signal on the transmit line is strongest at the transmission end. A signal on the receive line is also the weakest at this point. There is a greater likelihood of the coupled noise interfering with this weakened signal.

The formula for calculating NEXT in a cable is:

$$NEXT(dB) = 10\log\left(\frac{P_q}{P_a}\right)$$

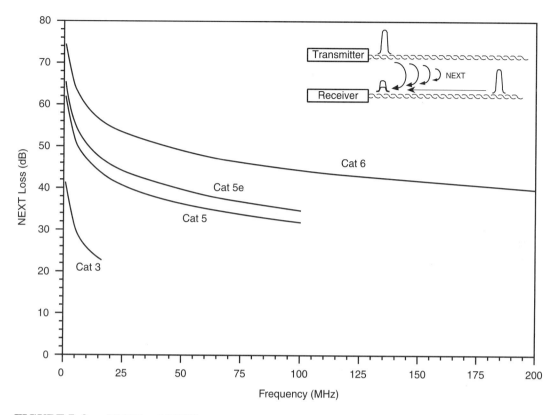

FIGURE 7–3a   NEXT and FEXT

where P_q is the noise power measured on the quiet line and P_a is the driving power on the active line. In other words, crosstalk is the ratio of crosstalk power to signal power, measured in decibels. Figure 7–3a graphs NEXT for various cable types.

FEXT

The original specifications for twisted-pair cable only specified NEXT. Most applications use only two pairs, one for transmitting and one for receiving. So you are only concerned with the effect of crosstalk from the transmitting line onto the receiving line at the near end. With some high-speed networks, notably Gigabit Ethernet, all four pairs of a cable are used to transmit and receive. Now FEXT becomes important. Consider again our active and quiet lines. We will now measure the crosstalk at the far end. The crosstalk will be the total of any crosstalk that occurs at any point along the line.

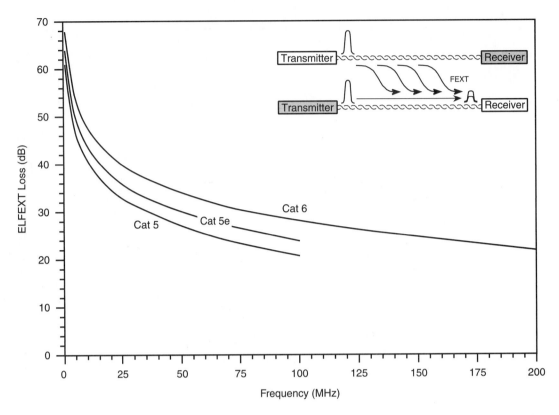

FIGURE 7–3b

While FEXT can be directly measured, cabling specifications do not use this value. Rather, they use a value for the *equal-level* FEXT or ELFEXT. ELFEXT is the ratio of the desired received signal to the undesired crosstalk at the far end:

$$\text{ELFEXT} = \text{Signal/FEXT}$$

Because the signal will be attenuated from the point at which it is launched at the near end and its arrival for measurement at the far end, ELFEXT can be calculated as the difference between FEXT loss and attenuation (both in dB):

$$\text{ELFEXT} = \text{FEXT} - \text{Signal Attenuation}$$

Because both the FEXT and the signal are attenuated as they travel, FEXT measurements are very low, lower than NEXT. But even the smaller power in the FEXT is important, because the signal is also much smaller. In NEXT loss, you compare how much smaller the

NEXT is compared to a relatively strong signal. In ELFEXT, you are comparing the loss to a weakened signal. Figure 7–3b graphs ELFEXT for various cable types.

Another consequence of ELFEXT is that crosstalk can occur at any point along the path from transmitter to receiver. NEXT occurs near the transmitter, but FEXT occurs anywhere. FEXT is the sum of all crosstalk that occurs along the entire path. Thus, it can be harder to isolate ELFEXT problems because a bad component or other problem can be anywhere in the path. With NEXT, you can usually assume that problems are occurring near the transmitting source.

There are three basic generalizations that we can make about noise and crosstalk:

Noise and crosstalk become more severe and harder to deal with at higher frequencies than at lower frequencies.

Noise and crosstalk exist in every copper-cable system. The trick is to minimize them to such an extent that they can be ignored.

An industrial setting can be very electrically noisy.

POWER-SUM NEXT AND ELFEXT

Standard NEXT and ELFEXT measurements simply compare the crosstalk from one pair to another pair. Only two pairs are involved. For many applications, this makes sense. An application like 10-Mb/s Ethernet only uses two pairs, one for transmitting and one for receiving. Thus, pair-to-pair crosstalk is of interest.

In *power-sum* NEXT and ELFEXT, we assume that energy is coupled from all other lines in the cable onto a single quiet line. Gigabit Ethernet uses all four pairs of a cable to carry a signal. A power-sum NEXT or ELFEXT measurement energizes three pairs and measures crosstalk on the fourth pair. We are comparing multipair-to-pair crosstalk. Another instance where power-sum measurements are important is in backbone or zone cables, which have more than four pairs. One popular type of cable uses 25 pairs in a single jacket. The power-sum test will energize 24 pairs and measure the crosstalk into the remaining quiet pair. Ideally, you should individually test all 25 pairs as the quiet line, so power-sum measurements can be quite involved.

In any crosstalk test, the area of greatest concern is the worst-case measurement. If the worst case passes, your cable passes. If the worst case fails, the cable fails.

Attenuation-to-Crosstalk Ratio

Think for a moment about signals and crosstalk. If a given level of crosstalk is coupled onto the cable, its relationship to the signal depends on the signal's power. If the signal is strong, then the crosstalk is only a small percentage of the signal. If the signal is weak, highly attenuated after traveling down the cable, then the crosstalk is a greater percentage of the signal. If the coupled crosstalk is .5 V and the signal is 5 V, the noise is 10% of the signal. If the signal is attenuated to 2.5 V, the noise is now 20% of the signal.

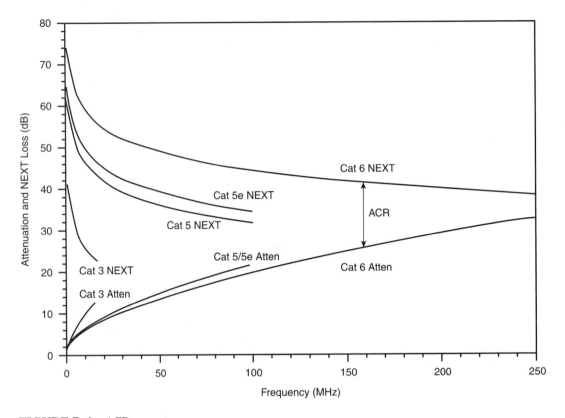

FIGURE 7–4   ACR

The attenuation-to-crosstalk ratio (ACR) is a figure of merit that accounts for the relative strength of the signal. Figure 7–4 shows the relationship between NEXT and attenuation. The distance that separates them defines the ACR. ACR generally decreases at higher frequencies. At 10 MHz for category 5 cable, NEXT loss is 47 dB and attenuation is 6.5 dB. The ACR is 31.5 dB. At 100 MHz, where NEXT is lowered to 32 dB and attenuation increases to 22 dB, the ACR is reduced to 11 dB.

To improve ACR, you can either reduce the crosstalk or lower the attenuation—or preferably do both.

The Decibel

NEXT and FEXT loss are measured in decibels. The decibel is used to compare two powers, currents, or voltages. In some cases, it compares the power going into the circuit to the power coming out the other end. In the case of crosstalk, it compares the power on one line to the power appearing on another line.

The basic equations for the decibel are:

$$dB = 20 \log_{10} (V_1/V_2)$$

$$dB = 20 \log_{10} (I_1/I_2)$$

$$dB = 10 \log_{10} (P_1/P_2)$$

where V is voltage, I is current, and P is power. The reason the decibel is a logarithmic ratio is to allow great variations to be expressed in small units. Each increase of 10 dB represents an order of magnitude ($\times$ 10) change in the ratio.

What does the decibel mean to NEXT, for example? Suppose we have 5 V on the driven cable. A NEXT of 20 dB means that 0.5 V appears on the quiet cable. The 20-dB means that the crosstalk voltage is 90% below the driving voltage. If the NEXT is 40 dB, the crosstalk is 99% below the driving voltage. Therefore, the higher the number, the greater the ratio and the lower the energy coupled onto the quiet line as crosstalk.

In dealing with decibels, it is important to keep in mind whether you are looking for a high number or a low number. With NEXT or FEXT, a higher number is better. It indicates that very little energy is coupled onto the quiet line. With the attenuation of a cable, a lower number is better: It means very little signal power has been lost. A high NEXT or FEXT number indicates better performance—that is, lower crosstalk.

Here is a general rule for understanding the decibel better. With power measurements, a loss of 3 dB means the power is cut in half. If you started with 1 mW, you now have 0.5 mW or 500 μW. Similarly, a gain of 3 dB means that the power is doubled: The 1 mW is now 2 mW. For voltage or current measurements, it is 6 dB that means the voltage or current is doubled or halved. (See the dB equations above: The ratio is multiplied by 10 for power and 20 for voltage and current. This difference also accounts for the difference between 3 dB or 6 dB.) In industrial cabling, we almost always are dealing with power, so the 3-dB figure is the one to remember.

With power, the decibel works out like this:

Gain: 20 dB = 20 times the power
10 dB = 10 times the power
3 dB = double the power

Loss: −3 dB = 50% power loss, 50% remaining
−6 dB = 75% power loss; 25% remaining
−10 dB = 90% power loss; 10% remaining
−20 dB = 99% power loss; 1% remaining
−30 dB = 99.9 power loss; 0.1% remaining
−40 dB = 99.9% power loss; 0.01% remaining

Characteristic	Better performance in dB
Crosstalk (NEXT/ELFEXT)	Higher number
ACR	Higher number
Structural Return Loss	Higher number
Cable Attenuation	Lower number
Connector Insertion Loss	Lower number

FIGURE 7–5   Putting the decibel into perspective

At only 20 dB, 99% of the power is lost. If you have 1 mW of original power, at 20 dB, you have only 10 μW remaining. You have dropped the power by two orders of magnitude. Each 10 dB represents an order of magnitude in gain or loss. Notice that while the equation will return a negative number when calculating loss, it is typical to omit the negative sign when speaking of loss. Thus –30 dB and a loss of 30 dB are the same.

Figure 7–5 is a table showing common cable characteristics that are measured in decibels and whether a higher number or lower number indicates better performance.

Another decibel unit you will often come across is the dBm, which means *decibels in reference to 1 milliwatt*. This simply means that 1 mW is used as a constant in the ratio

$$dBm = 10 \log_{10} (p/1 \text{ mW})$$

The dBm unit is widely used in fiber optics to simplify calculations of power in the system. A –10 dBm figure equates to 100 μW of optical power. dBm is preferred over microwatts because it allows losses (in decibels) to be directly added and subtracted. If your cable has a loss of 2 dB, and the connectors account for another 1 dB, then the output power is –13 dBm, or 50 μW.

Impedance

There is one other characteristic that can affect noise and signal transmission quality: impedance. Every cable has a property known as *characteristic impedance.* Characteristic impedance is determined by the geometry of the conductors and the dielectric constant of the materials separating them. For the coaxial cables we will be discussing next, determining characteristic impedance is easy since the cables have a uniform and regular geometry. In other cases, such as twisted-pair cables discussed later in this chapter, characteristic impedance is harder to determine because the geometry is more irregular. Figure 7–6 shows an example of how characteristic impedance can be determined in simple geometries.

The most efficient transfer of electrical energy occurs in a system where all parts have the same characteristic impedance. If a signal traveling down a transmission path meets a change

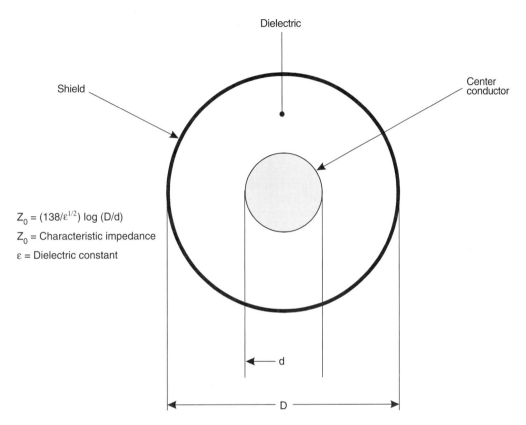

Dielectric

Shield

Center conductor

$Z_0 = (138/\varepsilon^{1/2}) \log (D/d)$

Z_0 = Characteristic impedance

ε = Dielectric constant

d

D

FIGURE 7–6 Characteristic impedance is determined by the geometry and dielectric constant of the cable.

in impedance, a portion of the energy will be reflected back toward the source. This reflected energy is also noise; that is, it is unwanted energy in the system that can distort signals.

How much energy is reflected depends on the severity of the impedance mismatch. If the mismatch is large in terms of the difference in ohms, a greater amount of energy will be reflected.

A component's electrical dimensions are an important concept in understanding signal transmission. Electrical length is compared to the wavelength of the signal traveling through the components. Figure 7–7 shows the relationship between a signal's frequency and its wavelength. A 10-MHz signal, for example, has a wavelength of 30 meters, while a 100-MHz signal has a wavelength an order of magnitude shorter—3 meters. If a component is much smaller than the wavelength, it is electrically insignificant. The signal does not "see" the component.

It is important to remember that this relationship between the wavelength of the signal and the physical size of components applies only to characteristic impedance and reflections.

Frequency	Wavelength (meters)
1 kHz	300,000
10 kHz	30,000
100 kHz	3000
1 MHz	300
10 MHz	30
100 MHz	3
1 GHz	0.3
$f = c/\lambda$ $\lambda = c/f$	

FIGURE 7–7   Frequency versus wavelength

Crosstalk and attenuation, on the other hand, are not similarly affected by a component's electrical dimensions.

The significance of a component's size can only be seen in comparison to the wavelength of the signal passing through it. As the signal speed increases, the wavelength decreases and the component becomes electrically larger. Therefore at very high speeds, even a small component becomes electrically significant.

Again, the most important thing to remember about impedance is this:

The most efficient transfer of energy occurs when all parts of a circuit have the same characteristic impedance.

Impedance mismatches are of greater concern at higher frequencies than at lower frequencies. The higher the frequency, the more important impedance matching becomes.

The reason is that at high frequencies, components tend to be electrically shorter and therefore more electrically significant than at lower frequencies.

The task of cable design is to devise cables that transmit signals as noise-free as practical. Once signal transmission quality is achieved, such other issues as cost and ease of use can be addressed.

COAXIAL CABLE

Coaxial cable offers the best high-frequency performance of all copper cables. Figure 7–8 shows the basic structure of a coaxial cable: center conductor, dielectric, shield, and jacket.

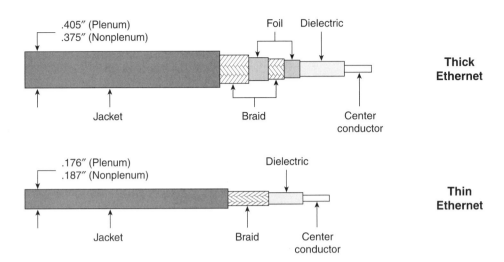

FIGURE 7–8   Basic construction of coaxial cable for Ethernet

These components are coaxial: They share the same center axis. As was shown earlier in Figure 7–6, the characteristic impedance of a coaxial cable is determined by three things: diameter of the center conductor, diameter of the shield, and the dielectric constant of the material separating them.

A coaxial cable is a closed transmission line. The shield serves to confine the energy within the cable. While we often think of a wire as carrying electrons through the conductor, it is equally important to think about the electromagnetic energy propagating down the wire. This electromagnetic energy is the magnetic and electric fields generated by the movement of the electrons. These fields travel on the outside of the center conductor, in the dielectric, and on the inside of the outer conductor. This system is called *closed* because the shield serves to keep the energy trapped within the cable.

In a coaxial cable, crosstalk is of very little concern. The shield works well at keeping the energy inside from radiating out or coupling to adjacent cables. Conversely, it protects the signal by keeping outside noise from getting in. Still, the problem remains of noise and signal distortion that results from impedance mismatches. This reflected energy, to a great degree, is also trapped within the cable.

But the shield is not perfect. Some shields are woven from very thin wires. Such braided shields provide coverage on the order of 85% to 95%. There are still gaps from energy to enter or exit. Other shields are foils, similar to aluminum foils. While these tend to provide 100% coverage, they are thin enough for some of the energy to penetrate. As a result, many cables use multiple layers of braid and foil.

Characteristic Impedance

Coaxial cables for Ethernet have a characteristic impedance of 50 ohms. CATV and other video systems typically use 75-ohm cable.

Termination

A coaxial cable must be terminated in its characteristic impedance. Some networks use a bus structure in which neither end of the cable necessarily connects to a device. A special connector called a *terminator* is placed on the cable end to absorb all the energy reaching it. This prevents reflections. If the cable is not terminated, the cable end represents an infinite impedance that will reflect nearly all the energy. If the shield and center conductor are shorted, the signal will also reverse and travel down the cable as noise. So termination is essential to proper network operation. Always be sure that the terminator has the correct resistance value. Terminating a 50-ohm cable with a 75-ohm terminator still causes reflections.

When Xerox first developed Ethernet, it used a large, sturdy, almost unwieldy coaxial cable with a diameter of either 0.405" for plenum applications or 0.375" for nonplenum applications. The cable uses a four-layer shield: braid, foil, braid, foil.

The cable has a 50-ohm (±2 ohms) characteristic impedance and uses N-type connectors. Its large diameter makes the cable stiff and, therefore, hard to work with.

By 1988, a new version of Ethernet had been developed that used a thinner, more flexible, and less-expensive coaxial cable. This 50-ohm cable, RG-58A/U, uses a stranded center conductor and a single-layer braided or foil shield. This version of Ethernet became popularly known as *thinnet* or *cheapernet*. The earlier version, in turn, was called *thicknet*. Thinnet cables use BNC connectors.

TWISTED-PAIR CABLE

Twisted-pair cable is the basic networking cable today, the workhorse that connects more Ethernet equipment than any other. Twisted-pair cables are so named because each pair of wires is twisted around one another, as shown in Figure 7–9. Some cables (Figure 7–9b) use

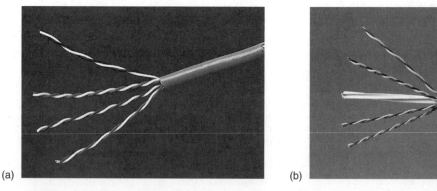

(a)   (b)

FIGURE 7–9   Basic twisted-pair construction: (a) cable without center spline, (b) cable with center spline
(© Avaya Communication. Used by permission.)

a plastic center spline to separate the pairs and achieve higher performance. Twisted-pair cables help reduce crosstalk and noise susceptibility in two ways.

First, the twists reduce magnetic coupling between pairs. The twists reduce inductive coupling between pairs. Inductive coupling is caused by the expanding and contracting magnetic fields caused by a signal through the wire. The twists create coupling among the two wires in the pair such that opposing electromagnetic fields are canceled. A greater twist ratio—the number of twists per unit length—brings greater coupling and less chance of crosstalk to nearby pairs.

The optimum twist rate involves a tradeoff between crosstalk and other cable characteristics. More twists lower crosstalk, but increase attenuation, propagation delay, and costs. High-quality category 5 cables have about one to three twists per inch. For best results, the twist length should vary significantly from pair to pair.

Second, networks using twisted-pair wiring use balanced transmission. The balanced circuit uses two wires to carry the signal. The conductors and all circuits connected to them have the same impedance with respect to ground and other conductors. Ground serves as a reference potential, rather than as a signal return.

Each conductor in a balanced line carries a signal (and noise) that is of equal potential with the other conductor, but of opposite polarity. If one conductor carries a 2.5-V signal, the other conductor carries a –2.5-V signal. The receiver detects only the difference between the two conductors—in this case, 5 V. If a +1 V noise is introduced onto each conductor, the resulting signal is 3.5 V and –1.5 V. The receiver still sees a 5 V difference. A perfectly balanced system is difficult to achieve, so some difference in noise levels may appear between the two conductors. Still, the effect is to cancel the noise.

Categories of Twisted-Pair Cable

As interest in twisted-pair cable for network applications grew, cable manufacturers developed better versions, capable of carrying higher data rates. Over the years, several categories or grades of cable have been developed, as summarized in Figure 7–10. The TIA/EIA has published specifications for categories 3 through 6 in its standard for commercial building cabling, TIA/EIA-568. Categories 1 and 2 were never standardized by a industry organization, although there was widespread understanding throughout the industry regarding them. The most recent version of 568 dropped category 4 in recognition that it was a marketing failure; nobody used it. Notice that the main distinction in the cables is the highest frequency they are specified to operate at.

Here is an important thing to remember about cable specifications: Frequency and bit rate are not the same. Do not think that a cable specified for 100 MHz can carry bit rates only up to 100 Mb/s. There are many ways to encode the 1s and 0s of digital data. One important aspect for high-speed networks is to encode the data in a manner that minimizes the frequency content of the signal. Gigabit Ethernet, for example, uses a complex method that requires considerable processing power at both the transmission and receiving end.

Category	Maximum frequency	Remarks
1	Not rated	Not specified by TIA/EIA
2	64 kHz	Not specified by TIA/EIA
3	16 MHz	Declining in popularity
4	20 MHz	Obsolete
5	100 MHz	The workhorse cable
5e	100 MHz	Enhanced category 5
6	250 MHz	High-end UTP: the emerging favorite

FIGURE 7–10   Categories of cable

No sooner had category 5 cable been standardized that many vendors brought out a "better than Cat 5" category 5 cable called enhanced category 5 or category 5e. Category 5e represents more than mere specmanship on the part of competitors. Gigabit Ethernet operates by dividing the signal among the four pairs of the cable—transmitting 250 Mb/s on each pair. In addition, it allows full duplex operation—transmission simultaneously in both directions. The original category 5 specification did not envision this application. While category 5 cable meets the needs of Gigabit Ethernet, it sometimes does so only marginally. For example, the original category 5 cable did not include a specification for FEXT or power-sum FEXT, which are important parameters to running full duplex over four pairs of cable. So, for most commercial applications, category 5e is a better choice than category 5.

Conventional wisdom would indicate that category 6 cable is better than category 5e cable, and 5e is better than 5. Generally this is true. But in industrial application, category 5 or 5e may be preferable to category 6. The reason is that category 6 cable, for example, can carry higher frequencies better (see Figures 7–3 and 7–4). This capability also makes category 6 cable more susceptible to the high-frequency noise from arc welders, motors, and other noise sources. Category 5 may be the better choice.

Industrial Cable: Category 5i

More recently, *industrial* category 5 cable—category 5i—has been developed in an attempt to achieve most of the characteristics of category 5e, but with an extended temperature range and a tougher jacket that is more resistant to abrasion, chemicals, and solvents.

What about category 6? If category 5e is undesirable because of its high-frequency characteristics, category 6 is even worse since it is even more susceptible to noise.

While conventional wisdom suggests it is best to put in the highest category of cable you can afford, factory applications show conventional wisdom is not always right. Category 5i is the best choice in noisy environments. You can carefully evaluate the electrical environment of the application and determine if industrial grade cabling is required or not.

Important note: As this book is being written, standards for category 5i cable are still under development. Although it appears that the standards will decide on a UTP cable with performance similar to category 5e, the committee may decide otherwise as engineering studies (and politics) progress. Proposals also have been made to model 5i cable more closely to category 5 or category 6. Other proponents favor shielded cable over unshielded. Thus, while our description seems to follow the prevailing opinion, the cable's performance requirements, once standardized, may differ. Perhaps the important thing to note here is that there will be a category 5i cable and it will be based (tightly or loosely) on an existing category of cable.

Unshielded versus Shielded Cable

The simplest of twisted-pair cable is unshielded twisted-pair cable or UTP. Each conductor has its own plastic insulation. The four pairs then have an outer jacket. In most office applications, UTP is preferred for the simple reason that it is the easiest to work with, the cheapest, and the most flexible. And even while twisted-pair cable has very good noise immunity, UTP can still be sensitive to external electrical noise.

To provide better noise immunity, shielded twisted-pair cable is often used. There are two types. The first is what we will call classic shielded twisted-pair cable (STP). With STP, each pair has its own shield—plus an overshield surrounding all the pairs. While this construction gives excellent shielding, it is cumbersome, expensive, and not widely used. STP differs from UTP in that it has a 150-ohm impedance. STP was originally developed by IBM for its Token Ring network. As with the network, it has fallen into disfavor.

A more convenient alternative is screened twisted-pair cable (ScTP), which has only an outer shield surrounding all the pairs. While there are some minor differences in electrical specifications, you can think of ScTP as simply being a shielded version of UTP—same categories, same NEXT, same 100-ohm impedance, and so forth. ScTP is an attractive choice in noisy environments. To realize the benefits of ScTP shielding, you must use shielded connectors and connect the cables to shielded ports on the equipment. The essential duty of the shield is to pick up the noise and conduct it to ground. Therefore, you must provide a continuous path for the noise. Shielded ports are quite common on Ethernet equipment.

UTP Connectors

Ethernet uses RJ-45 modular plugs and jacks (Figure 7–11) as the preferred connector. The RJ-45 is an 8-position version of the connectors used on a telephone. They are inexpensive, easy to use, and provide excellent performance—at least in office applications. Like cables, Ethernet connectors are available in categories. To achieve category 5 performance, you need category 5 cable *and* category 5 connectors. Put another way, you can only expect the performance of the lowest level of component. If you use category 3 connectors on category 5e cable, you cannot expect better than category 3 performance (you might get better, but do not expect it).

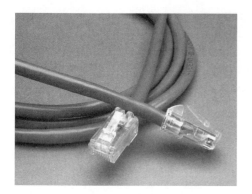

FIGURE 7–11 Modular plugs and jacks
(© Avaya Communication. Used by permission.)

While most networks applications only use two pairs (one pair carrying signals in one direction and one pair carrying them in the other direction), it is wise to use four-pair cable. One of the original ideas behind this thinking was to lay cable that could accommodate different types of networks. In a cabling system using all four pairs and standard wiring rules, Token Ring and Ethernet operate on different pairs. The practical consequence is that you could wire once and run different networks—otherwise you need to reconfigure the wiring to accommodate the type of network being used. In other cases, it is possible to run telephone line on two pairs and Ethernet on two pairs. In factory applications, it is possible to run power on unused pairs or to use them for other purposes.

Figure 7–12 shows the two accepted wiring patterns for an RJ-45 connector. These are called T568A and T568B. The only difference is that pairs 2 and 3 are swapped. As long as the entire system is wired with the same pattern, the system will work.

The problem with RJ-45 connectors is that they were not designed for the factory. They are not built to withstand lots of vibration, to provide a sealed interface against moisture or splashing, or to be robust enough to endure mechanical abuse. At least two alternative connectors have been suggested for Ethernet applications.

Figure 7–13 shows a sealed RJ-45 connector. A threaded boot sitting over the connector provides both sealing from moisture, gases, and liquids and better stability from vibrations. It provides IP 67 sealing and reduces the effects of vibration. The threaded coupling is much more resistant to vibration, supporting the contacts sufficiently to support them and prevent undue wear.

The second connector, shown in Figure 7–14, is the cylindrical metal-shell M12 connector. These sturdy connectors were designed for rugged use.

Figure 7–15 shows an example of an Ethernet switch equipped with the overmolded IP67 RJ-45 connectors. The device is waterproof and can be used in a wide range of applications where even "standard" DIN rail-mounted switches would be inappropriate. It can even be built right into machines that will need to be washed down or otherwise subjected to water or other liquids.

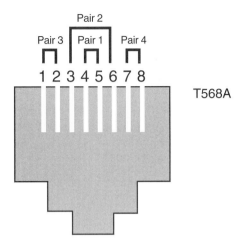
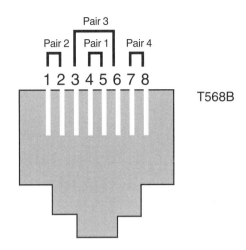

FIGURE 7–12   RJ-45 wiring patterns

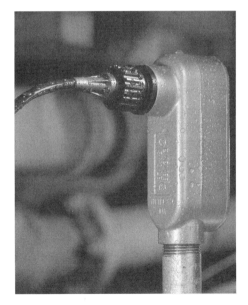

FIGURE 7–13   Industrial Ethernet connectors: sealed RJ-45
(Courtesy of Woodhead Industries)

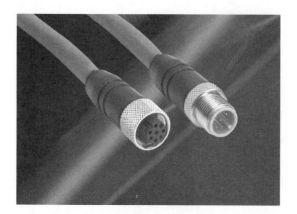

FIGURE 7–14   Industrial Ethernet connectors: M12 cylindrical and M12-to-RJ45 adapter
(Courtesy of Lumberg)

FIGURE 7–15   Ethernet switch, with sealed connector interfaces and capable of being mounted on
a machine
(Courtesy of Woodhead Industries)

FIBER OPTICS

The optical fiber offers an attractive alternative to copper cabling for high-speed communications. An optical fiber is a hair-thin strand of glass that carries signals as light. Depending on the type of fiber, the information-carrying capacity can range to many gigabits per second over distances of hundreds of kilometers—well beyond the practical capabilities of copper cable.

The optical fiber consists of two basic parts: the inner light-carrying core and the outer cladding. The cladding serves to help confine the light inside the core. There are two main types of optical fibers: multimode and single mode. In either case, fibers typically have a cladding diameter of 125 μm. A multimode fiber has a core diameter of 50 or 62.5 μm, rather large by fiber standards. This large diameter provides many paths for rays of light and thereby limits the bandwidth of the fiber to several hundred megahertz per kilometer.

A single-mode fiber has an exceedingly small core—only about 9 μm. This permits only one path for the light and results in an extremely high bandwidth—tens or even hundreds of gigahertz per kilometer. While copper twisted-pair cable is designed to carry Fast Ethernet and Gigabit Ethernet over distances of 100 meters, optical fibers extend transmission distances to 300 meters for multimode cable and 40 kilometers for single-mode fiber.

Fiber-optic technology brings four great benefits to the transfer of information:

1. *High bandwidth.* The information-carrying capacity of a fiber is greater than that of twisted-pair cable.
2. *Low attenuation.* The lower attenuation of a fiber permits longer transmission distances before the signal is diminished too far. Equally important, attenuation in a fiber does not increase with the signal frequency as it does with copper cable. For example, category 5 UTP has an attenuation of 6.5 dB at 10 MHz and 22 dB at 100 MHz over 100 meters. A multimode fiber has an attenuation of 3.5 dB over 1000 meters—or .35 over 100 meters at 10 MHz and 100 MHz.
3. *EMI immunity.* A fiber is neither a source nor receiver of electromagnetic inference. You simply do not have to worry where you run the cable from the perspective of electrical noise.
4. *Safety.* A fiber is not a spark or fire hazard. It can be safely used in factory settings.

Fiber Variables: Or What You Need to Know

There are several factors that influence the transmission capabilities of an optical fiber. The main ones are the wavelength of light, the type of source, and the type of fiber.

WAVELENGTH

Colors of light are distinguished by their wavelength. The light used in most optical fiber systems is infrared light, meaning you can not see it. In fiber optics, the wavelength of light is expressed in nanometers (nm), which is a billionth of a meter. The attenuation of light in an optical fiber depends on the wavelength of light traveling through it. Light is attenuated more at 850 nm than at 1300 nm. Three low-loss "windows" are most often defined for use in fiber-optic systems: 850 nm, 1300 nm, and 1550 nm. The first two are widely used by networks, including Ethernet. The 1550-nm region is used mainly for high-speed, long-distance telecommunication applications, including the Internet and telephony. 10 Gigabit Ethernet is the only flavor that support the 1550-nm band, and because 10 Gigabit Ethernet is not of major concern in factory applications, we can safely ignore this window.

The 850-nm region is called the *short-wavelength* band; the 1300-nm region is the *long-wavelength* band. Thus, 1000BASE-SX refers to Gigabit Ethernet running at 850 nm (the *S* referring to *short* wavelength), and 1000BASE-LX operates at 1300 nm. (Watch out here. This is the way network people look at short and long wavelengths. Telecom people refer to 1300 nm as the short wavelength and 1550 nm as the long wavelength. So, 1300 nm is in either the short band or the long band, depending on who is talking.)

Why the different wavelengths? Cost is one important reason. LEDs and lasers operating at 850 nm are cheaper to make than those at the longer wavelengths. A simple 10BASE-FL LED can cost a few dollars. A high-performance laser in telecom applications can run thousands of dollars.

Attenuation in a fiber varies with the wavelength. For example, a multimode fiber can have an attenuation of 3.5 dB/km at 850 nm and 1.5 dB/km at 1300 nm. A single-mode fiber has an attenuation of 1 dB/km or less.

LIGHT SOURCE

An electrical signal is converted into an optical signal by either an LED or a laser. An LED is inexpensive, but it emits a wide band of light. For example, a typical spread of light in a 10BASE-FL LED is 40 nm. What this means is that if the nominal wavelength is 850 nm, the light emitted actually covers a 40-nm range from 830 to 870 nm. For many reasons, the wider the spectrum of light emitted by the source, the lower the bandwidth of the signal the fiber can carry. The practical bandwidth of a fiber depends in part on the nature of the light going through it. A narrow output band from the source is better.

Enter the laser, which has an output spectrum of less than 0.1 nm in high-end telecom applications. For practical purposes, this means a single wavelength. In Ethernet applications, the wavelength spectrum of a laser is around 1 nm to 3 nm. The laser offers three main benefits over the LED:

1. *Narrow spectrum.* As mentioned, the wavelength spectrum is 3 nm and often under 0.1 nm less.
2. *High speed.* Lasers can be operated very quickly, giving data rates of 2.5 Gb/s or more.
3. *Narrow, intense output beam.* An LED emits light in a wide pattern, so that only a small portion is coupled into the fiber and transmitted. Many LED-based transmitters

use a lens mechanism to focus the light into the fiber. The laser's beam is narrowly focused, with minimal spreading to allow very efficient coupling of light into the fiber. The small core of a single-mode fiber requires use of fiber.

In general, we think of lasers as being used with single-mode fibers and LEDs with multimode fibers. The VCSEL (for vertical-cavity surface-emitting laser) is a special type of low-cost laser that is compatible with both types of fibers. It can operate faster than an LED, bringing gigabit speeds to multimode fibers. And it allows transmission distances of several hundred meters. In other words, VCSELs extend the data-carrying capacity and transmission distances of multimode fibers. Plus they can be used with single-mode fibers for shorter runs (say, under 10 km).

In most factory applications, single-mode fibers find use in backbone applications, running between switches or, more often, between buildings. Because multimode fibers are more compatible with the plant floor, we will concentrate on them.

Ethernet and Fiber

Ethernet operates at 10, 100, and 1000 Mb/s over fiber-optic cable and supports distances up to 500 meters over multimode fiber, 3000 meters over single-mode fiber for 10-Mb/s and Fast Ethernet, and up to 40 km for gigabit Ethernet. As a general rule, cable runs of multimode fiber are typically limited to 300 meters. The reason is that this distance allows any type of Ethernet equipment to be connected to the fiber and run. Just a few years ago, multimode fiber with a 62.5-μm core diameter was the overwhelming choice, specified as the fiber of choice. This core diameter was chosen over a 50-μm core as offering the best trade-off of adequate bandwidth and a core size that made coupling of light easier. The march of technology, however, has made the industry look again at the 50/125 fiber (with the first number being the core diameter and the second number the cladding diameter). The 50/125-μm fiber has a higher bandwidth and can carry signals a longer distance. And the VCSEL is well matched to couple light into the smaller core.

Figure 7–16 summarizes the characteristics of typical fibers. Notice that, like copper cable, there are several "grades" of multimode cable, distinguished by attenuation and bandwidth. The TIA-EIA-568 version is the specification in the standard for commercial buildings. However, these "vanilla" cables had trouble supporting Gigabit Ethernet over distances of 300 meters unless you use a special "conditioning" cable with it. Better cables overcome this problem and allow even further distances, with 500-meter capabilities being common.

Another important specification to notice in Figure 7–16 is the effect of the VCSEL laser on bandwidth. For the enhanced 50/125 fiber, the bandwidth achievable with a VCSEL is over four times that of an LED.

Notice that bandwidth is expressed in terms of a distance: 500 MHz-km. In general, bandwidth scales linearly with distance. If the bandwidth is specified as 500 MHz-km, the fiber will support 1000 MHz over half that distance (500 meters) or 250 MHz over double the distance (2 km).

Fiber Size	Fiber "Grade"	Bandwidth (MHz-km)		Attenuation (dB/km)	
		@ 850 nm	@1300 nm	@ 850 nm	@1300 nm
50/125-µm fiber	TIA/EIA-568B	500	500	3.5	1.5
	Enhanced (A)	500 (LED) 2200 (laser)	500 (LED) 500 (laser)	3.5	1.5
62.5/125-µm fiber	TIA/EIA-568B	160	500	3.5	1.5
	Enhanced (A)	200	500	3.5	1.0
	Enhanced (B)	200	1000	3.5	1.0

FIGURE 7–16   Multimode fibers

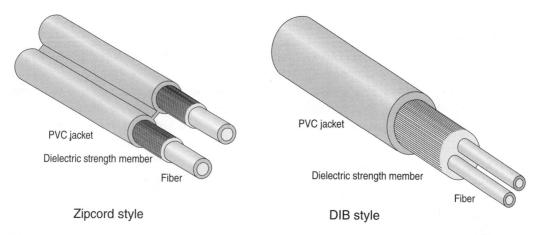

Zipcord style   DIB style

FIGURE 7–17   Common duplex cable constructions
(Courtesy of Corning Cable Systems)

Fiber Cable

The fiber, of course, must be cabled—enclosed within a protective structure. This usually includes strength members and an outer jacket. The most common strength member is Kevlar aramid yarn, which adds mechanical strength. During and after installation, strength members handle the tensile stresses applied to the cable so that the fiber is not damaged. Steel and fiberglass rods are also used as strength members in multifiber bundles. Figure 7–17 shows typical fiber constructions.

The first layer of protection is the buffer, available in loose and tight styles. The tight buffer construction as a plastic layer applied directly to the fiber. This buffer is typically 250 or 900 µm thick. Widely used indoor applications, tight-buffer cables provide great crush and impact resistance, but lower isolation of the fiber from extreme changes in temperature.

The breakout cable is a special form of tight buffered cable—cables within a cable. It consists of several individual tight-buffered cables (fiber, buffer, strength member, and outer jacket) with an outer jacket and additional strength members. Useful for both riser and horizontal applications, breakout cables can simplify routing and installation of connectors.

In the loose-buffer cable, the fiber lays within a plastic tube with an inside diameter many times that of the fiber. A tube can hold more than one fiber. The tube is usually filled with a gel material that serves to keep water out. Since the fiber floats within the tube, it is isolated from external stresses, including expansions and contractions of the cable due to temperature extremes. Widely used in outdoor applications, loose-buffer cables tend to have a larger bend radius, larger diameter, higher tensile strength, and lower crush and impact resistance.

Duplex cables are used for patch cables and many horizontal cables. For backbone and riser cables, cables with twelve or more fibers are recommended.

For factory applications, heavy-duty cables with abrasion- and contaminant-resistant jackets are available.

FIBER-OPTIC CONNECTORS

Fiber-optic connectors differ significantly from electrical connectors. Connectors for both copper and fiber have a similar function: to carry signal power across a junction with as little power loss and distortion as possible. But the difference between electrical signals and optical signals means that the connectors are very different. Where the electrical connector forms a gastight, low-resistance interface, the key to an optical connector is alignment. The light-carrying cores of the fibers must be very accurately aligned. Misalignment means that optical energy will be lost as light crosses the junction.

Alignment of the two fibers is a formidable task. Not only are the dimensions quite small—a core diameter of 50 μm for multimode fibers and 9 μm for single-mode fibers, tolerances complicate the matter. One hopes that the core is perfectly round and perfectly centered within a perfectly round cladding—and that the cladding is 125 μm, not 125.5 or 126 μm. The same with the connector dimensions. Unfortunately some variations creep into any manufacturing process. These variations represent misalignment that can increase the loss of optical power.

Even so, both fiber and connector manufacturers have evolved processes to create tightly toleranced components. Some years ago, the goal for insertion loss was 1.5 dB. Then 1 dB. Today's connectors offer loss of around 0.1 to 0.3 dB; cabling standards typically call for a loss of 0.75 dB or less.

Loss is measured as insertion loss. Insertion loss is the loss of optical power contributed by adding a connector to a line. Consider a length of optical fiber. Light is launched into one end, and the output power at the other end is measured. Next the fiber is cut in two, terminated with fiber-optic connectors, and the power out is again measured. The difference between the first and second measure is the insertion loss.

Most traditional connectors use a 2.5-mm ferrule as the fiber-alignment mechanism. A precision hole in the ferrule accurately positions the fiber. The two ferrules are brought together in a receptacle that precisely aligns the two ferrules through a tight friction-fit. A new generation of connectors uses a smaller-diameter ferrule to significantly reduce the size of the connector.

Ferrules are typically made of ceramic, plastic, or stainless steel. Until recently, there were clear differences between the materials. Ceramic was more expensive and offered lower insertion loss. Plastic was the least-expensive material, but offered higher insertion loss. Stainless steel fell in between. Today the price and performance of the materials is much closer, so there is little difference. The main exception is single-mode applications, where ceramic is still the preferred material.

Epoxy has long been a staple in ensuring the fiber is properly held in the ferrule. And epoxy has long proven that it does its job well. But epoxy has also been seen as a necessary evil. It has several drawbacks. It is an extra step in the process, it is potentially messy, and it requires curing. Curing time can be shortened with a portable curing oven, but that means another piece of equipment to lug around. Finally, in working in a completed building, the possibility of spilling epoxy on an executive's carpet or furniture is an unpleasant prospect.

There are several ways to apply a connector to a fiber:

- Epoxy and polish
- No epoxy/polish
- No epoxy/no polish

The traditional method was to epoxy the fiber into the ferrule, cleave the end of the fiber flush with the ferrule, and finally polish the end face of the ferrule/fiber. The epoxy bonds the fiber to the ferrule; the polishing ensures a clean, flawless optical finish to maximize the transmission of light. There are many types of epoxy available today to meet different conditions, from curing conditions to application conditions. While this epoxy/polish method can result in the best terminations, it has some drawbacks:

- The epoxy is messy and (depending on the type of epoxy) may have a long curing time
- There is additional equipment (curing ovens or polishing machines) and consumables (polishing paper and epoxy) to purchase
- It is the least productive, requiring the most time and labor

Another method is to use a mechanical clamping mechanism inside the connector to firmly hold the fiber. The fiber must be cleaved and polished.

Yet another method uses a small stub of fiber inserted into the connector in the factory. This is part of the factory-polished end face. The fiber to be terminated is inserted into the connector and spliced to the fiber stub. Neither epoxy nor polishing is required. The method is fast and efficient. Although a no-epoxy/no-polish termination typically has a slightly higher insertion loss than the best epoxy/polish termination, the difference is minimal and still yields a termination well within the requirements of factory equipment.

Types of Connectors

There is no shortage of types of fiber-optic connectors. For convenience, we can define three generations. The first generation included many different approaches. The most successful was the so-called SMA connector, which used the same form factor as the SMA coaxial connector. While some older automation equipment still uses the SMA connector, it has become largely obsolete.

The second generation used the 2.5-mm ceramic ferrule discussed previously. For factory applications, the two most successful were the ST and SC connectors. Both have been widely used in Ethernet equipment.

The third generation are the so-called small-form-factor connectors, which are about half the size of ST and SC connectors.

Figure 7–18 shows typical fiber-optic connectors.

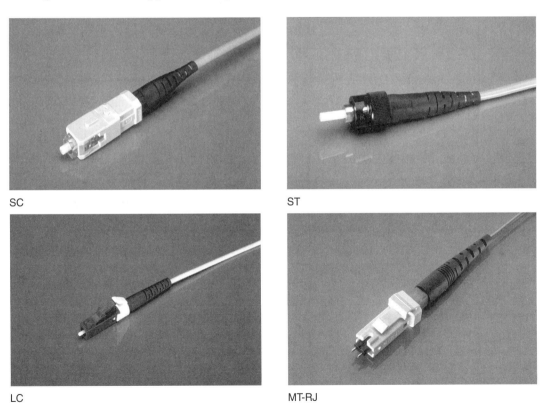

SC

ST

LC

MT-RJ

FIGURE 7–18   Fiber-optic connectors
(Courtesy of Corning Cable Systems)

SC Connector

Nippon Telephone and Telegraph (NTT) originally designed the SC connector for telephony applications. It has subsequently become the preferred general-purpose connector, being widely used in Ethernet and factory applications. Three characteristics make it popular.

Its design protects the ferrule from damage. The connector body surrounds the ferrule so that only a small part sticks out. It uses a push-pull engagement mechanism. Finger access is easier and requires less all-around space than a connector with a rotating engagement mechanism like threaded or bayonet-style coupling nuts. This permits closer spacings and higher density designs in patch panels, hubs, and so forth. In addition, the rigid body eliminates intermittency problems caused by accidental pulling on the cable at the rear of the connector. Connectors are easily snapped together to form multifiber connectors. Joining two connectors to form a duplex connector simplifies installation, use, and troubleshooting of the premises cabling plant. Since all links require two fibers, each carrying signals in opposite directions, a duplex cable assembly obviously makes sense. If you have ten point-to-point links, it is easier to deal with ten duplex cables rather than twenty individual cables.

ST Connectors

Invented by AT&T, ST connectors were the connectors of choice during the late 1980s and early 90s, before being replaced by the SC. The ST uses a bayonet coupling that requires only a quarter turn to engage or disengage the connector.

Small-Form-Factor Connectors

Small-form-factor—or SFF for short—connectors are considerably smaller, offering a receptacle about the same size as a modular jack. They typically use a similar push-to-release latch similar to RJ-45 connectors. A major advantage is that they can achieve the same port density as that of a RJ-45 connector. Consider a network switch having twenty-four copper side-by-side RJ-45 ports. In the same space, you can only fit twelve SC duplex ports. (Remember you need a duplex fiber port, one for the incoming signal and one for the outgoing signal; when we talk of optical ports, we mean duplex ones.) With an SFF connector, you can achieve twenty-four duplex ports—the same as for copper. This is an extreme space-saving advantage for SFF connectors. Not only does it save space in real-world applications, it simplifies the design of equipment. Consider an enterprise switch with plug-in board. It becomes possible for the equipment designer to design a board with either twelve copper ports or twelve fiber ports.

While much existing equipment today offers SC or ST ports, the advantages of SFF connectors mean the market will migrate to these.

MT-RJ CONNECTOR

The MT-RJ connector was designed by AMP (now Tyco Electronics), Siecor (now Corning Cable Systems), and USConec as an alternative to RJ-45 modular telephone connectors. The

MT-RJ uses a small rectangular ferrule that holds two fibers. It also uses a press-to-release latch quite similar to that found on the modular jack to make its operation familiar.

Most fiber-optic connectors use two plug connectors mated through a coupling adapter. While the MT-RJ can be used in this manner, it also offers a true plug-jack interconnection. Designed for very simple termination of fiber, the jack contains two fiber stubs in a polished ferrule. The back end of the stub fibers are in a mechanical splice. To terminate the fiber-optic cable, the fibers are cleaved, slid into the splice, and the splice is closed to form a low-loss interconnection.

LC Connector

The LC connector was designed by Lucent Technologies as a smaller connector for premises cabling and telecommunications applications. Based on a 125-mm ceramic ferrule, the LC uses an RJ-45-style housing with a push-to-release latch that provides an audible click when the connector is mated in an adapter. Available in both duplex and simplex configurations, the connector doubles the density over existing ST and SC designs. The connectors are color-coded beige for multimode and blue for single mode.

There are two things to mention about SFF connectors in particular and fiber-optic connectors in general. First, we are concerned with the connector as an interface. Transceivers should also be available with the interface to allow the interface to be used throughout the system. SFF transceivers are the key to higher port densities in equipment.

IS FIBER OPTIC CABLE READY FOR INDUSTRY?

Over many years and in many different applications, copper-based media has been engineered to handle the rugged industrial environment. In much less time, optical fiber media is being adapted for demanding applications.

Rugged Fiber

It is common to assume that because an optical fiber is glass, it is fragile, and certainly too fragile for the factory. Actually, the optical fiber—and the fiber-optic cable in particular—can be quite rugged. An optical fiber is stronger than steel of the same diameter. The main drawback to fiber is that it is brittle and cannot be sharply bent: it will snap. It can be bent in loops, and the cable serves as a rugged exterior to protect the glass.

Ruggedized Connectors

As with copper connectors, fiber connectors have been "ruggedized" for factory applications. Versions of fiber-optic connectors similar to the RJ-45 connectors in Figure 7–13 have been shown, but (as of this writing) are not commercially available.

The second thing to remember is that connector choice does not have to be an either/or choice. You can put one type of connector on one end of the cable and another type of connector on the other end. Suppose, for example, you cable your factory with the LC interface. Does this mean your network equipment must also feature the LC? What if you buy an Ethernet switch with an MT-RJ interface? No problem. You simply need a cable with a duplex LC connector on one and an MT-RJ connector on the other.

Even so, it might be easier and more convenient to have an end-to-end system with the same interface. It will be easier to buy patch cables off the shelf, without special ordering this connector here and that connector there.

SAFETY AND FIBERS

There are two special safety considerations to be aware of when working with fiber-optic cables.

1. Never look in the end of a fiber unless you are *absolutely* sure there is no light being transmitted through the fiber. Some light is powerful enough to cause injury or blindness. Most industrial Ethernet applications do not use light sources powerful enough—they are considered eye safe. But it is better to err on the side of safety.
2. Beware of tiny slivers of fiber you may produce. These slivers are sharp and can easily penetrate the skin. They are irritating and often hard to remove.

Cleanliness

The second most important rule (after safety) in handling and using optical fibers is to keep the connections clean. The importance of cleanliness cannot be overstated. Dust, dirt, and other contaminants have two detrimental effects. First, they can increase the insertion loss until they are wiped away. Second, they can permanently scratch the end surface of the fiber, which then requires re-polishing or a new termination. Two methods of cleaning are "canned" air and special wipes. In a pinch, isopropyl alcohol and a lint-free cloth will do.

UTP OR FIBER—OR BOTH?

Factory applications support both copper UTP and optical fibers. To a large degree, the fiber is a superior transmission media in an electrically noisy factory. It offers higher bandwidth, longer transmission distances, and immunity to electromagnetic interference. Yet copper is more familiar, supports data rates up to Gigabit Ethernet over distances of 100 meters, and is less costly in acquisition costs.

In many cases, UTP will be used at the device level and fiber will be used for backbone and higher-level connections. Fast Ethernet provides sufficient bandwidth at the device level. Even at the higher control and information levels, Fast Ethernet will be widely deployed, except in cases where bandwidth demands make a move to Gigabit Ethernet advisable. Gigabit prices have dropped faster than most observers forecasted, and Gigabit links for backbone and server connections are widely available. Since many backbone connections extend beyond the 100 meters that UTP can carry, multimode fiber is the choice for runs from up to 300 to 500 meters and single-mode fiber for distances beyond 500 meters.

Some Notes on Practical Cabling

- Use Category 5i UTP on the factory floor in environments with low electrical noise.
- Use ScTP in electrically noisy environments. For maximum noise immunity, consider using fiber.
- Ground a shielded cable only at one end. Grounding at both ends allows noise-inducing ground loops. A cable grounded at both ends provides a complete circuit loop through the conductor and shield. If the ground system's voltage varies, it can set up noise currents in the cable. Grounding at one end prevents ground loops.
- Use a sealed connector where vibration, dust, or moisture might degrade a standard connector.
- Consider conduit when you are not sure the cable will be adequately protected. Use a sealed bulkhead connector. A metal conduit provides excellent EMI protection.
- Install four-pair cable in most cases. This will ensure the maximum application flexibility: for example, power can be added on the unused pairs.
- Make sure all cables are properly terminated. Incorrectly installing the connector is a major cause of performance degradation. A category 5 cable with a improperly applied connector might yield only category 3 performance.
- Install network equipment in enclosures to protect them from moisture and dust. If necessary use bulkhead adapters to mate the inner cable with the outside cable. This will often serve more reliably than a sealed access hole.
- Do not subject cables to sharp bends. Use gentle radii instead. For UTP, the radius should be four times the cable's outer diameter—and no less than 1 inch. For fiber, the minimum is also four times the cable diameter.
- Keep copper cable away from high-voltage power sources and power cables. High-voltage is 480 V and above. If the cable is installed in open or nonmetallic pathways, use the following guidelines:
    Power cable of less than 2 kVA: 5-inch minimum separation
    Power cable from 2 kVA to 5 kVA: 12-inch minimum separation
    Power cable over 5 kVA: 24-inch minimum separation
  If the UTP is installed in a shielded pathway, you can cut the separation distance in half.
- Carefully identify and mark each cable with a unique identifier. Be sure to carefully document each cable and the device it is connected to at each end.

SUMMARY

- The three types of cable used by Ethernet are coaxial cable, twisted-pair cable, and fiber-optic cable.
- Twisted-pair cable is the most widely used cable with Ethernet.
- Twisted pair cables are available in unshielded (UTP), screened (ScTP), and shielded (STP) versions. While not popular in commercial applications, ScTP and STP offers better noise performance in factories.
- Fiber-optic cable is preferred for high bandwidths, in longer distances (over 100 meters), or in noisy environments.
- Attenuation is loss of signal power.
- Crosstalk is the coupling of energy from one conductor to another.
- NEXT measures crosstalk at the near end of the cable; FEXT measures it at the far end.
- ACR is the difference between attenuation and NEXT at a given frequency.
- UTP cables are "ranked" by category, which is a measure of comparative performance.
- Category 5i is an industrial-grade UTP cable.
- The maximum recommended length for UTP cable is 100 meters.
- Fiber-optic cable offers higher bandwidth, longer distances, and better noise immunity than UTP.
- The two main types of optical fiber are multimode and single mode.
- Multimode cable is recommended for distances up to 300 meters; single-mode fiber carries signals thousands of kilometers.
- UTP cable uses RJ-45 connectors in commercial applications; industrial applications use the RJ-45, an industrial RJ-45, or the M12 cylindrical connector.
- Fiber-optic cable uses the SC, ST, MT-RJ, and LC, among others. Seal versions are being developed.

? REVIEW QUESTIONS

1. What is the difference between NEXT and FEXT?
2. Why is ACR a useful metric for evaluating the performance of UTP cable?
3. What is the maximum frequency for category 3, 5, and 6 UTP cables?
4. What benefit does STP cable bring to industrial applications?
5. What are two main differences between category 5i and category 5e cable?
6. Why is category 5i cable a better choice than category 6 cable for industrial applications?
7. What are the two major parts of an optical fiber? Which carries the light? Which serves to guide and reflect the light?
8. What type of fiber is used for longer distances between buildings?
9. What type of fiber is used on the factory floor to connect equipment over shorter distances?
10. What is the disadvantage of epoxy in terminating a fiber-optic connector? What is the advantage of epoxy?

Wireless Ethernet

8

OBJECTIVES

After reading this chapter, you will be able to:

- Discuss three advantages of wireless Ethernet
- Distinguish between IEEE802.3 and 802.11
- Define CSMA/CA
- List the two types of spread-spectrum signaling used by wireless Ethernet and tell how they differ from one another
- List the four main flavors of wireless Ethernet and their speeds
- Define three ways that Bluetooth differs from wireless Ethernet

WHY WIRELESS?

Our discussion of Ethernet has emphasized 802.3 Ethernet running over UTP and optical-fiber cables. Wireless Ethernet offers an alternative in application where cabling is difficult or impractical.

Wireless networking has two main advantages:

- It untethers a worker from the network. The worker is free to roam about, accessing part and process specifications. A wireless connection with real-time sensing permits problems to be remotely diagnosed. Plant equipment can be maintained and monitored by a mobile engineer. Handheld scanning or programming devices can be used anywhere, with a network link being maintained wirelessly. This permits real-time communications, rather than periodic dumps of information.

In a wired LAN the station's logical address also implies a physical location. With a wireless network, it does not. The station's MAC address tells you nothing about its physical location.

- It allows a network link where cabling may be impractical. In a hostile environment— where you certainly do not want to run copper and perhaps not fiber, wireless links eliminate the need for cables.

The 802.11 Standard and Wi-Fi

Wireless Ethernet is the responsibility of the 802.11 committee of the IEEE. The 802.11a standard concerns operation in the 5 GHz band, and 802.11b concerns the 2.4 GHz band. Recently a separate organization, the Wireless Ethernet Compatibility Alliance (WECA), has promoted the IEEE 802.11b standard under the name **Wi-Fi,** which stands for **wi**reless fidelity. There is no need for confusion: Wi-Fi is simply marketing nomenclature for 802.11b to allow WECA to certify products as interoperable. In practice, any Wi-Fi product can use any brand of access point with any other brand of client hardware, just as long as they meet the Wi-Fi (802.11b) standard.

Why is the IEEE's wireless Ethernet standards not handled by the same 802.3 committee that specifies cabled Ethernet? The short answer is that wireless Ethernet uses a different media-access method. Recall that the 802.3 standard is for a network using CSMA/CD; that is, it is based on carrier sense, multiple access with collision detection to prevent multiple transmissions simultaneously.

While wireless Ethernet still uses the carrier-sense, multiple-access idea, it uses another technique called collision avoidance in place of collision detection: Thus the 802.11 standard covers CSMA/CA.

Collision avoidance works like this:

- When a station is ready to send, it listens to the network for traffic. If it detects a transmission, it backs off for a random time.
- The station first sends a short ready-to-send (RTS) packet that contains information about the length of the packet.
- The receiving station responds with a clear-to-send (CTS) packet. On hearing the CTS, all other stations stop transmitting until the station has successfully transmitted its message.
- When it receives the CTS packet, the requesting station will transmit its message.
- If the packet is correctly received (as determined by the CRC), the receiving station sends an acknowledgment (ACK) to the transmitter.

Since the CTS sequence also notifies other stations (other than the requesting transmitter) that a transmission is desired, they know a transmission is to follow and so they know not to transmit. This method significantly avoids collisions, although it does not eliminate them. Most collisions, however, occur with the RTS message. Since the RTS message is very short, the likelihood of collision is smaller.

One reason for this change is that a transmitting station cannot detect collisions. When a station is transmitting, the only thing its receiver will detect is the power from the transmitter. The transmitter drowns out all other signals.

AD HOC OR INFRASTRUCTURE NETWORKS

There are two basic approaches to building a wireless network, depending on whether devices communicate directly or through an intermediary access point.

Simple, Wireless Network

Figure 8–1 shows a simple wireless network. Two PLCs and a supervisory computer, each equipped with a wireless transceiver, communicate directly with one another. There is no hub or other intermediary device such as is used with a cabled network. This arrangement is called an *ad-hoc network.*

Infrastructure Approach

The second method is the infrastructure approach, shown in Figure 8–2. Here stations communicate through a central hublike access point. The access point can also connect to the wired LAN. The area covered by a single access point is called the basic service set (BSS). By connecting access points over a wired LAN, you can achieve an extended service set (ESS). With ESS, stations using different access points can be in contact.

ESS INFRASTRUCTURE
A nice feature of infrastructure networks using access points is the ability to roam. Suppose a mobile technician is connected to access point A. Eventually, he will move out of the range of A. But as he does so, his access will automatically be passed to access point B. When a station first enters an ESS, it associates itself with the access points with which it establishes

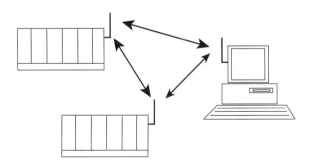

FIGURE 8–1   Simple ad hoc wireless network

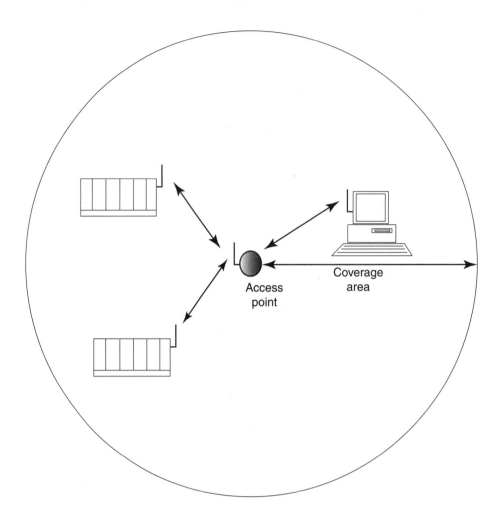

FIGURE 8–2   BSS infrastructure wireless system

the best communications (based on error rates and data rate). Periodically, it will scan all channels to see if another access point offers a better connection. As the user moves about, the connection with one access point will deteriorate while the connection with another access point will improve. The station will drop the first access point and associate itself with the second access point. The advantage is that the user can roam wherever there is an access point and stay connected to the network. At least as of this writing, roaming was poorly defined by 802.11. Each vendor tends to devise its own protocol to support roaming. As long as all equipment is from the same vendor, roaming works well. If you mix access points from several vendors, you may have problems. Some vendors have joined to develop an Inter-Access Point Protocol to ensure interoperability. Figure 8–3 shows an ESS network.

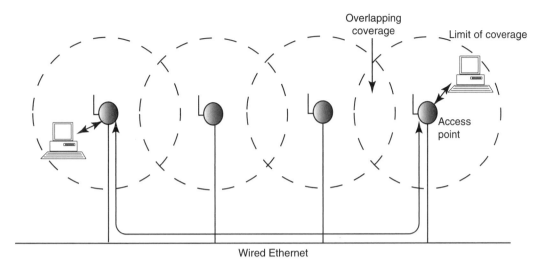

FIGURE 8–3   ESS infrastructure wireless network

FLAVORS OF WIRELESS ETHERNET

Like all networks, wireless Ethernet is an evolving standard. The main versions are summarized in Figure 8–4. Each version has a maximum data rate and fallback rates. Because the quality of transmission between station and access point can vary with the distance between them, the details of the physical environment, and other factors, a station may fall back to a slower speed to achieve reliable communication.

Standard	Data rates (Mb/s)		Radio band (GHz)	Radio technique
	Max	Fallback		
802.11	2	1	2.4	FHSS/DSSS
802.11a	54	48, 36, 24,* 18, 12,* 9, 6*	5	DSSS
802.11b	11	5.5, 2, 1	2.4	DSSS
802.11g	22	11, 5.5, 2, 1	5	DSSS
*Mandatory fallback rates that must be implemented for compliance. Other rates may or may not be supported by a particular device.				

FIGURE 8–4   Flavors of wireless Ethernet

Standards

Versions 802.11 at 2 Mb/s and 802.11b at 11 Mb/s are the most widely used wireless versions today, although the higher speed versions may become popular if demands for speed grow. In many factory applications, high-speed links are unnecessary, so 11 Mb/s and lower rates are adequate.

Distance

The distance between stations or between station and access point is very dependent on the physical environment in which they are used. The propagation characteristics of radio signals are complex and unpredictable. Even small changes in the position or direction of a wireless station or access point can dramatically improve or degrade signal transmissions or reception. You could locate a 11-Mb/s wireless station at point A and achieve 11 Mb/s. Move a few inches this way and the speed could fall to 5.5 Mb/s. Move it a foot another way and speeds could fall back to 2 Mb/s.

Distance can play an important role in signal strength and possible data rates. Generally, signal strength falls off with distance. A slower signal can pack more energy into each bit, so that signal quality can be improved by slowing down the signaling rate. Figure 8–5 shows the idea of practical signal rates versus distance for an 802.11b 11-Mb/s network. In the real world, you do not always achieve the full data rate—fallback is common. For example, one widely used wireless Ethernet device is specified to operate at 1750 ft at 1 Mb/s or 525 ft at 11 Mb/s. This is for open conditions (without obstructions). In a more representative semi-open application, distances are 375 ft at 1 Mb/s and 165 ft at 11 Mb/s. In a closed environment, distances can reduce to 80 ft at 11 Mb/s. The point is that speed and application environment play a significant role in potential transmission distances.

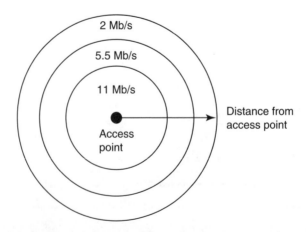

FIGURE 8–5   Distance versus speed in a wireless network

The maximum distance specified by vendors is for an open-space environment, such as a field where there are no obstructions. Most vendors will also specify typical real-world operating distances, typically for an office environment. It is often good to perform a site survey which will show the signal intensity throughout the physical environment. This allows the vendor to recommend maximum distances and even location limitations for access points and stations.

Bridging

Wireless Ethernet can also be bridged. A common reason for bridging is to connect two LAN segments that are separated by a wider distance than can be achieved otherwise. For example, a bridge can be used to connect two buildings. While this link can also be done with fiber, in some cases it is not easy or convenient to do so. Suppose that two buildings are separated by a major highway, river, or other physical feature that makes running a cable difficult. A wireless bridge forms an easy way to link the two buildings. In some cases, these bridges could be the only wireless portion of an otherwise wired LAN. Wireless bridges work as a point-to-point or point-to-multipoint link. They do not provide coverage for communicating with stations; they connect to an access point or to a wired LAN segment.

Some access points can handle more than one channel at a time. Essentially, they contain multiple antennas and radios to allow simultaneous communication with more than one station. Stations simply have to make sure they don't hop onto the same frequency at the same time.

RADIO TECHNIQUES

Figure 8–4 lists two types of radio techniques used in wireless LANs: frequency-hopping spread spectrum (FHSS) and direct-sequence spread spectrum (DSSS). Wireless Ethernet operates in one of two unlicensed radio bands: 2.4 GHz and 5 GHz. The 2.4-GHz band is called the ISM band, for the industrial, scientific, and medical equipment that uses the band. In the United States and Europe, the band extends from 2.4000 GHz to 2.4835 GHz—a width of 83.5 MHz. The less widely used 5-GHz band extends over a 300-MHz in three ranges, each with a different allowable output power:

Low Band: 5.15 to 5.25 GHz, 50 mW
Middle Band: 5.25 to 5.35 GHz, 250 mW
High Band: 5.725 to 5.825 GHz, 1 W

In the United States, all three bands are available for unlicensed use, meaning they can be used by a wide range of applications. In Europe, only the low and middle bands are available.

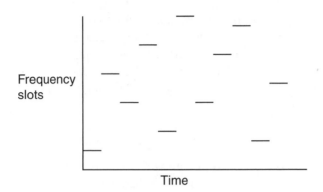

FIGURE 8–6   Frequency-hopping spread spectrum

If two applications transmit on the same narrow frequency, the signals can interfere with one another. These interfering sources include not only wireless Ethernet stations, but also any other instrumentation, scientific, or medical devices operating in the area. The trick in wireless Ethernet is to avoid putting all your eggs in one basket or, in this case, call your signal in one narrowband frequency. Equally important, the Ethernet transmission should not interfere with other devices. Wireless Ethernet accomplishes this by spreading the signals out over a wide band of the spectrum, a technique called *spread spectrum*. There are two approaches to spread spectrum.

Frequency-Hopping Spread Spectrum

Frequency-hopping spread spectrum uses over 75 frequencies. A transmitter transmits on one frequency for a brief period, hops to another frequency for a second brief period, hops to a third, and so on (Figure 8–6). The pattern is random, but predictable: The receiver must hop in the same pattern to receive the signal. The duration of transmission on any given frequency is 400 ms or less. Any interference is of short duration since it will occur at only one of the frequencies. The information is simply retransmitted (in all probability on another frequency). To an unintended receiver tuned to one of these frequencies, the transmission will appear as a short burst of impulse noise.

One drawback to FHSS is that the need for both the transmitter and receiver to hop among frequencies and to remain synchronized as they do so limits the practical data rate to 2 Mb/s. While FHSS is a very usable technique, it is falling out of favor with wireless Ethernet. Only the original version of 802.11 specified FHSS and a revised version allows both FHSS and DSSS for 2-Mb/s systems.

Direct-Sequence Spread Spectrum

Direct-sequence spread spectrum uses a different approach. It generates a high-bandwidth bit pattern called a *chip* or *chip code* (Figure 8–7). (The term *chip* is sort of clever. A bit is bro-

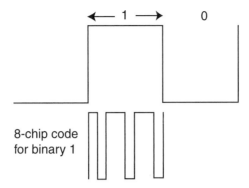

FIGURE 8–7   DSSS generates a redundant bit pattern called a chip, which allows data to be recovered if one or more chip bits are damaged.

ken into chips!) The chip contains redundant information about the bit of information. If one or more bits of the chip get damaged or interfered with during transmission, the receiver can use sophisticated statistical techniques to recover the information without the need for retransmission. The longer the chip code, the easier it is to recover the information and the more interference that can be tolerated before the information is lost.

The encoded (chipped) data is interleaved on fifty-two separate subcarriers. The effect of this interleaving is to spread the signal over a wide spectrum. Instead of concentrating the power in a narrow frequency (as does FHSS, although it hops from one power-packed frequency to another), DSSS spreads the power of a wide range of frequencies. To an unintended receiver, the DSSS signals looks like low-power, wideband noise that is easily ignored.

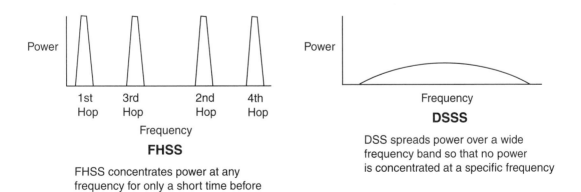

FIGURE 8–8   FHSS versus DSSS

BLUETOOTH

Bluetooth is a complementary wireless networking scheme aimed at so-called personal area networks (PANs). A PAN is a network with a geographic coverage considerably smaller than a LAN. It operates similar to wireless Ethernet in that it uses the 2.4-GHz ISM band and FHSS, using twenty-three or seventy-nine hopping frequencies. However, it is different in that it operates on a master-slave arrangement. *Piconet* may be the preferred term for Bluetooth in industrial applications, since a factory is not really dealing with personal and consumer electronics.

Origin

Bluetooth was originally conceived as a low-cost, short-distance cable-replacement technology in computers and consumer electronics. For example, printer, scanners, digital cameras, and all sorts of other peripherals can be linked to a PC by Bluetooth. Similarly, a cell phone can be linked to a headset by Bluetooth. An important feature is that a Bluetooth host can immediately recognize when a device is added or removed from the network. The host must discover devices added to the network and understand basic requirements for communication with the device. Different types of devices have profiles that define its functions and how the host is to treat it. The first set of profiles for Bluetooth equipment included cordless telephones, intercoms, serial ports, headsets, dial-up networking (wireless modems), faxes, and LAN access, as well as generic devices.

Bluetooth operates at a maximum 1 Mb/s over short distances—about 30 feet or so. In factory application, it can be used to create, for example, a machine network for linking I/Os and controls in a complex piece of machinery, with uplinks to Ethernet. Bluetooth can operate in either point-to-point or point-to-multipoint configurations. In multipoint operation, one device serves as a master that controls channel access. Other devices act as slaves. Up to seven slaves can be active, although additional slaves can be attached and inactive. These inactive devices are referred to as *parked*. This point-to-multipoint configuration is termed a *piconet*.

Piconets and Scatternets

Several piconets can operate in an overlapping area of coverage—in an arrangement called a *scatternet*. Each piconet has its own master, but a slave can participate in more than one piconet: that is, it can serve more than one master. In addition, a master of one piconet can be a slave in another piconet. Figure 8–9 shows these different types of Bluetooth operation.

Bluetooth is flexible. Piconets can communicate with one another, and masters and slaves can reverse their relationships. The master can become a slave and the slave a master.

Point-to-point

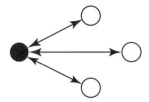

Piconet
(Point-to-multipoint)

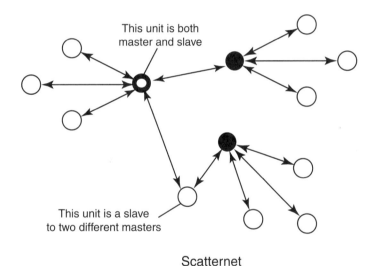

Scatternet
(Multiple piconets)

FIGURE 8–9   Types of Bluetooth operations

As this book is written, Bluetooth is still a young technology, generating more excitement than working products. Still, Bluetooth in factory applications can be used to create a machine area network, that is, to create a wireless fieldbuslike network on a machine. It also offers a low-cost way to provide a wireless interface to bar code readers and handheld terminals used for calibration. As Bluetooth and other technologies drive down the cost of RF/wireless applications, RF tags are becoming popular for identifying items as they travel through a factory or warehouse.

OTHER WIRELESS TECHNOLOGIES

Other radio technologies can be used in factory applications, mostly for communications beyond distances covered by Ethernet. While they generally fall outside the scope of this book, we will briefly mention them. Point-to-point radio links and even satellite links provide means of connecting widely separated points. Satellite links, using global positioning system data, are used to track the progress of inventory or even to locate lost or stolen goods. Radio links along pipelines or in large processing plants can create links by stations separated by tens or hundreds of miles. While many of these applications (such as GPS) are specialized, others may be replaced by Ethernet and the Internet. Many applications that previously required expensive bought or leased capabilities can often use the Internet to accomplish the same purpose more simply and economically.

SUMMARY

- Wireless Ethernet provides advantages where cabling is difficult.
- Wireless Ethernet uses a different access method from standard 802.3 Ethernet.
- There are several versions of wireless Ethernet, operating in different frequency bands, at different data rates, and using different radio techniques.
- The two main frequency bands used are 2.4 GHz and 5 GHz.
- The radio techniques are frequency-hopping spread spectrum and direct-sequence spread spectrum.
- Bluetooth is another wireless technology used to create personal area networks over short distances.

? REVIEW QUESTIONS

1. Why are wireless Ethernet standards written by the 802.11 committee instead of the 802.3 committee?
2. What access method does Ethernet use?
3. What is another term used for *personal area networks* that may apply better to industrial applications?
4. What is the maximum data rate for wireless Ethernet operating with FHSS?
5. What is the difference between an ad hoc and an infrastructure wireless network?
6. If a user can roam among two or more access points and stay connected, does the network offer BSS or ESS services?
7. What is the effect of distance on a wireless Ethernet network?
8. How does Bluetooth differ from Ethernet in terms of transmission distances?
9. How does Bluetooth differ from Ethernet in terms of access method?
10. What is a scatternet?

PART THREE

Protocols for Network, Transport, and Application Layers

Building the Industrial Ethernet Protocol Stack: Fieldbus Is Being Framed

9

OBJECTIVES

After reading this chapter, you will be able to:

- Understand why application-layer protocols are crucial to the success of industrial Ethernet and manufacturing-to-business interconnections.
- Explain how various control and automation protocols combine their distinctive application-layer protocols with Ethernet and TCP/IP.
- Understand why industrial Ethernet protocols are incompatible with each other.
- Appreciate how recently offered solutions resolve these application-layer incompatibilities.

Part Two covered Ethernet's physical and data-link layers, plus some of the functions of IP in the network layer. That is the foundation for the iE transport- and application-layer mechanisms discussed here.

In theory, Ethernet along with TCP/IP is an ideal platform on which to build an alternative to proprietary fieldbus protocols. In practice, however, different parties have used that platform to create their own proprietary application-layer protocols. So instead of a single, open Ethernet standard for factory automation and control, today there are several incompatible iE versions.

This chapter examines the core network-, transport-, and application-layer technologies used in nine iE protocols:

1. Modbus TCP
2. Ethernet/IP
3. Fieldbus HSE
4. PROFInet
5. iDA-RTPS
6. Interbus
7. MMS Ethernet

   8. ADS-net

   9. .NET for Manufacturing (DCOM)

If you are more interested in the specific features of these protocols rather than their general mechanisms, go to Chapter 10, which examines the individual communication stack of each one. Also in Chapter 10, an integrating application-layer protocol is introduced—OPC DX—that promises to enable data exchange between different iE protocols.

PLAYING WITH A STACKED DECK

Before examining the mechanisms used by these competing protocols, the reader is excused for asking: "But wasn't the whole point of developing an iE standard precisely to eliminate this kind of confusion?"

Control companies who developed these protocols do not necessarily have a problem with confusion. Instead of pursuing a dream of designing one open, universal protocol, they had more pragmatic concerns—to stay in business by engineering a solution for their particular equipment and for their customer's process, automation, and discrete control problems.

You could say it is a matter of "different strokes for different folks." Or you could call it the **Tower of Babel effect.** For years, each company remained loyal to its own technology. Today even with a common Ethernet TCP/IP platform, their allegiances translate into a cacophony of different application-layer standards. And that is why users, control system integrators, and book writers exert so much effort over the problem of communication between different protocols and devices.

Compare the situation with the telephone system. The phone line is analogous to Ethernet as the physical medium. The dial tone, ring, on-hook, and off-hook procedures are the TCP/IP transport mechanisms. The spoken language is the application-layer protocol. The underlying technologies work just fine for all languages, but if the language isn't spoken by the user, confusion results. The key is to use a language that meets the user's needs.

The Protocol Stack

Due to competing agendas and differing needs, a universal application layer was never really in the cards. The safe bet was that vendors would take advantage of Ethernet's open foundation to build their own distinctive protocol stack—which is what happened.

As mentioned in Chapter 1, the **communication** (or **protocol**) **stack** describes the hierarchical arrangement of rules that are followed during the network communication process. A single protocol is the set of rules agreed upon by two entities to ensure they cooperate when performing certain communication tasks; a protocol stack is the complete set of steps used to perform all the interactions necessary for communication (Figure 9–1).

Layer	Protocol	Standard
Transport Layer	UDP	RFC 788
	TCP	RFC 793, 761, 675
Network Layer	IP	RFC 791
	ICMP	RFC 792
	Broadcast	RFC 919, 922
	Subnet	RFC 950
Data-link Layer	Ethernet frame	RFC 894
	ARP	RFC 826
Physical Layer		IEEE 802.3

FIGURE 9–1   Standards followed by lower layers of the iE protocol stack

A protocol stack exists as software code that resides in the device's memory—embedded in a ROM chip or in an operating system. The code prepares the data for transmission by composing the proper headers, packet format, and contents and then initiating the communication process. At the receiving end, the program takes the packets off the wire. The first protocol in the stack divides and analyzes (parses) the headers, extracting and processing the data contained in the field of interest, and then passes the body up for handling according to the protocol in the next layer.

The iE Protocol Stack

The iE protocol stack is built on top of Ethernet's physical and data layers, plus the network, transport, and application layers. Different iE versions are distinguished by the choice of specific protocol functions used in these three layers:

1. *Network-layer choices.* May employ one of three types of network addresses: unicast, broadcast, and multicast.
2. *Transport-layer choices.* May employ either TCP, UDP, or both.
3. *Application-layer choice.* May employ a number of different protocols for application-to-application communication interactions and services, functions like user authentication, data formats, error handling, etc., needed by the application program.

Because the presentation layer (5) and session layer (6) are not defined by Internet protocols, they are not usually treated as separate parts of the iE stack, but are lumped into a general "upper layer." Furthermore, the so-called "user layer" exists outside the iE stack, because it describes the end-user's actual application.

User layer: Application program		
Request/reply channel	Message stream	Process-to-process communication
Transport/Network Layer: Host-to-host connection (TCP/IP, UDP/IP)		
Physical Layer: Network interface, etc.		

FIGURE 9–2   Three different services provided at the application layer

The Confusion Starts at the Top

The application layer is where the differences in iE protocols are most evident. That is because the application layer provides communication services specific to the device and application program.

In Chapter 1, the focus was on how an application-layer protocol (HTTP) and TCP/IP transmitted data as a reply to a request or as a message stream. But the application layer can also provide services that help run program processes (Figure 9–2).

The complex control schemes and data flows mentioned in Chapters 2 and 3 require sophisticated connections between application programs running on different devices. These requirements directly impact the design of the protocol stack, especially the application layer.

P2P MESSAGING
When writing an application program, a developer can use protocols for one of two basic communication models. In the point-to-point (P2P) model, information is exchanged between one sending application and one receiver—which is why P2P messaging is the basis for client/server communication.

PUB/SUB MESSAGING
The other model is the publish/subscribe (pub/sub) model, where a single sending application can transmit one message to several receiving applications. Pub/sub messaging is useful when a single message must reach a number of devices at the same time.

The choice of messaging model depends on how the application program is written.

Different Programs, Different Protocols

Application programs can be written using one of four programming approaches, each of which is predicated on a specific communication model (Figure 9–3).

Program design	Application-layer protocol design
Hard-coded	Serial communication Raw sockets
Modular	Socket programming

Object-oriented	RPCs
	ORBs:
	   DCOM
	   CORBA
	   Java RMI

Agent-based	Messaging Pub/sub

FIGURE 9–3 Relationship between programming approaches and application-layer protocols

HARD-CODED PROGRAMMING

Early application software was written as a single, integral unit of code, including the code for serial communication. Programs written as integral units are typical in legacy applications using master/slave serial communication or fieldbus protocol.

MODULAR PROGRAMMING

As programs became more complex and required extensive rewriting, modular programming techniques were developed that allow small units of code to be reused. This development roughly coincided with the rise of the Ethernet and a modular approach to writing communications programs known as **socket programming.**

OBJECT-ORIENTED PROGRAMMING

Early programs were based on complex logical statements that processed inputs into desired outputs. In contrast, object-oriented programs are constructed from entities that respond to events according to their own code segments, known as **methods.** Object-oriented programming makes it possible to dispense self-contained portions of executable code—known as **components**—onto remote devices that can do the processing without depending on a central processor. This is the idea behind distributed computing and distributed control. To get these components to interoperate across a network, object-oriented application-layer protocols are used, such as **CORBA (Common Object Request Broker Architecture), DCOM (Distributed Component Object Model),** or **Java RMI (Remote Method Invocation)** scripts.

AGENT-BASED PROGRAMMING

Like objects, **agents** are also comprised of self-contained code. But these objects are also self-monitoring and self-directed based on the individual rules and goals of the distributed logic. Consequently, they can initiate a whole set of interactions on their own, both issuing and responding to any message they choose. Agent-based and object-oriented programming are both used in distributed computing. Distributed applications employ the pub/sub form of messaging.

Caught in the Middleware

To allow applications to communicate over a network, several kinds of programming techniques are used—namely sockets, remote procedure calls (RPC), object request brokers (ORBs), and pub/sub—which are collectively known as **middleware.**

As the name implies, middleware is software that enables communication between applications. Functionally, middleware moves information from one program to another, allowing devices to exchange data regardless of their location or hardware type.

From a programming standpoint, middleware is the code that operates between the application program's code and the TCP/IP stack in the operating system's software (Figure 9–4).

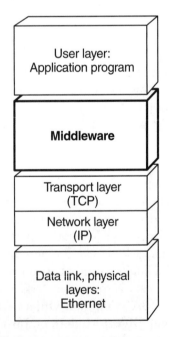

FIGURE 9–4   The place of middleware in the protocol stack

INTERPOCESS COMMUNICATION

In technical terms, middleware is employed in **interprocess communication (IPC)** to provide a set of interfaces or tools so processes in different programs can run concurrently to enable the exchange of data and parameters.

ENTERPRISE APPLICATIONS

The concept of middleware dates back to the 1980s when programmers developed ways to move data between mainframes, databases, and user terminals. Today, middleware is commonly used in distributed computing applications. In business enterprises, middleware is employed in **N-tier applications,** so called because the application spans many different levels. N-tier applications are often developed in conjunction with an **Enterprise Application Integration (EAI)** initiative where data is exchanged between different databases, servers, and ERP systems. Such an initiative may end up pulling data from a plant's MES or SCADA system—even down to PLCs.

As seen in Chapter 3, the business model, manufacturing system, and production strategy are intertwined with the application architecture and network topology.

The grand strategies of business information technology experts, the brilliant applications written by programmers, and the practical concerns of control engineers all collide in the middleware.

HOW MIDDLEWARE STACKS UP

For most automation and control companies, the choice of middleware has been dictated by the need to maintain the desirable characteristics of their legacy protocols and applications. Here is how IPC mechanisms and iE protocols correlate:

1. *Socket-based TCP/IP.* Used in Modbus/TCP
2. *RPC.* Used in PROFInet
3. *Object brokers/components.* Used in PROFInet and Ethernet/IP
4. *Message passing.* Used in Foundation Fieldbus HSE, iDA, Ethernet/IP, MMS TCP/IP, and ADS-net
5. *SOAP.* Used in .NET for Manufacturing (discussed in Chapter 10)

Each IPC method may differ in a number of features, such as connection type, relationship between devices, transport protocol, transaction technique, IP addressing, application coupling, packet sequencing, service level, overhead, transmission latency, data throughput, device assignment, and port range (Figure 9–5).

Function	Sockets TCP/IP	RPC	DCOM	Message passing
Connection	One-to-one	One-to-one	One-to-one	One-to-many
Relationship	Client/server or peer-to-peer	Client/server or peer-to-peer	Client/server or peer-to-peer	Producer/consumer
Transport	TCP	TCP or UDP	UDP	UDP
Transaction	Request/reply	Procedure call: Request/reply	Method call: Request/reply	Publish/subscribe
IP addressing	Unicast or broadcast	Unicast	Unicast	Multicast
Coupling	Tight	Tight	Tight	Loose
Sequence	Synchronous, blocking	Synchronous, blocking (threaded option available)	Synchronous Asynchronous	Asynchronous, nonblocking
Services	Low level	Low level	High level	High level
Overhead	High (TCP) Low (UDP)	High	High	Low
Latency	High (TCP) Low (UDP)	High	High	Low
Throughput	High	Low	Low	High
Device assignment	Manual	Manual	Managed	Managed
Port usage	Restricted range or dynamic	Port mapping via port 111	Dynamic, one port per process	Assigned range

FIGURE 9–5 Features of IPC mechanisms

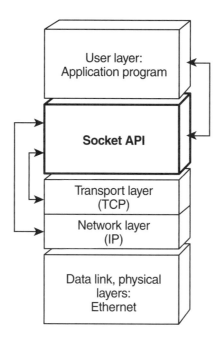

FIGURE 9–6   Socket in the application layer

Socket-Based TCP/IP: Used in Modbus TCP

Sockets are the primary tool used for interprocess communication between devices in a client/server relationship (Figure 9–6). Defined from a software program's point of view, a socket is the hook that an application uses to pull TCP (or UDP) into action so it can transmit data as if it were a normal input/output function, like reading or writing to a floppy disk.

From the network's point of view, a socket is one end point of a two-way communication link that binds the source and destination addresses with their respective port numbers (Figure 9–7).

GENERAL SOCKET FUNCTION

A server application uses a socket to maintain an open port—such as port 80—to listen for incoming requests. Thereafter, the process is analogous to dialing a multiextension phone system. Like a caller dialing the phone, the client sends a one-way command to establish a connection. If accepted, the server "answers the phone" by creating a new socket. But just like a call being routed to a specific extension, the socket binds the reply to a different port number so it can continue to listen on the original socket for additional incoming requests. On the client side, the server sends a command that assigns (sometimes randomly) a distinct port number to create a client socket, thus completing the connection. This set of sockets is created specifically to make the connection and torn down when the communication is finished.

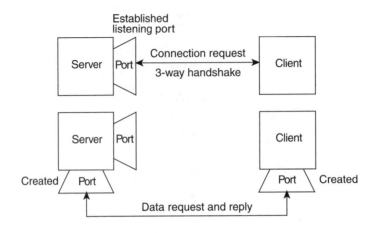

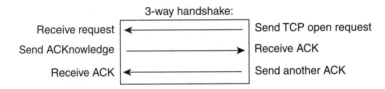

FIGURE 9–7   Socket function

There are two widely used socket types, **stream sockets** and **datagram sockets.** Stream sockets create a specific connection to handle a continuous number of packets, while datagram sockets use a connectionless method to send a packet of data in a predefined length. Stream sockets use TCP (transport control protocol), which has a reliable method of sequencing streams. Datagram sockets use **UDP** (variously named as **Unix, User,** or **Unreliable Datagram Protocol**).

After a socket reads the data from the application's memory buffer, it calls upon the transport protocol layer to determine the delivery mechanism—either TCP or UDP—and then binds the port number to the appropriate IP address—either unicast or multicast.

IP ADDRESSING FOR SOCKETS
Internet protocol version 4 (IPv4) performs four tasks that enable internetworking across dissimilar networks. In previous chapters we mentioned IP's error handling, routing, fragmentation, and addressing functions. Of interest here are the three address types used by sockets, namely, unicasting, broadcasting, and multicasting (Figure 9–8).

Unicast addressing.   The conventional method of networking connects two computers in a one-to-one relationship. Data goes from computer A to computer B by simply creating a packet with computer B's IP address, then the network delivers it. The details of IP addressing are discussed in Appendix B. Unicast addressing provides an individual address for each source and des-

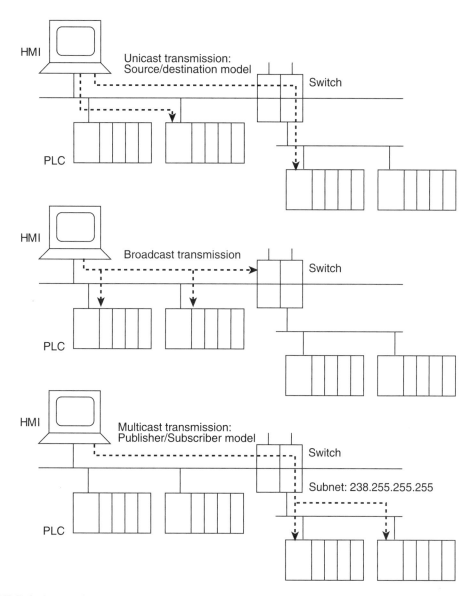

FIGURE 9–8 Unicast, broadcast, and multicast transmission

tination device to create a unique one-to-one connection. But what if the same information must be transmitted to a number of devices? Then the source must make a copy of the message, establish a unique connection, and send it to each destination one at a time. Where many identical messages are involved, unicast messaging wastes both the sender's and the network's resources.

Broadcast addressing.   A more efficient use of the sender's resources is to transmit one message and deliver it everywhere. With broadcast messaging, the IPv4 header can be addressed to reach every station in the local network domain (subnet). Or the header can contain an address that tells routers to forward the packet—a potentially troublesome feature that is ideally limited to legitimate uses like an address resolution protocol (ARP) broadcast. Broadcast addressing is not used by iE protocols.

Multicast addressing.   The most efficient way to get a single message to multiple nodes is to send it only to the interested group. Multicast addressing does this by using the unicast packet format, and by changing the IP header's destination address to a **class D IP address,** which signifies the packet should be routed to a member of a multicast group.

A node can join the group by being configured with an IP subnet address in the range reserved for multicasting: 224.0.0.1 to 239.255.255.255. Then it must advertise that fact by sending an **Internet group management protocol (IGMP)** report with the new address to network routers.

Multicast messaging also involves a change in the Ethernet MAC header. The first byte in the destination address is coded "01" to allow the network interface to quickly detect a multicast packet. Normally the field is coded "00," permitting the NIC to receive only packets that match its own hardware address.

Because the multicast mechanism does not require each individual device's explicit IP addresses to establish a connection, a single UDP packet can be transmitted to many hosts. If the packet meets a fork where it must travel to multicast group members on two or more networks, the packet is replicated by the router.

TCP SOCKET FUNCTION

A stream socket binds a port number to a unicast IP address. With this end point established, TCP divides the outgoing message into a continuous sequence of packets addressed specifically to that device. The end-user's application does not have to account for packet delivery; TCP takes care of the job.

A three-way handshake establishes the connection between the client and the server. Here is an example of a how an HTTP server uses TCP/IP to establish a communication session with a client (such a transmission may be preceded by an ARP broadcast as mentioned in Chapter 1):

1. The server opens a TCP socket that binds the IP address with port 80 to listen for HTTP requests. The client sends a request-to-connect datagram bound for port 80 at the server's destination address. The server's NIC recognizes its address and pulls the datagram off the wire; the packet is then processed by TCP/IP. The server opens another socket assigning its IP address to a new port number and sends a reply to the client. The client sends an acknowledge packet addressed to the new port number, concluding the three-way handshake that opens the connection.
2. The server's socket writes data from the application buffer and uses TCP/IP to form the reply. The initial packet contains a number indicating the quantity of packets in the message.

3. The server socket loops, writing data in packets until the entire message is sent.
4. The TCP client compares the number of packets with the quantity in the initial packet and acknowledges receipt of the entire message.
5. Sockets close—the connection is torn down.

Throughout the transaction, TCP maintains an orderly, reliable communication process. That is because when handling messages, TCP is **connection oriented**—that is, it keeps the connection open until all packets in the message are delivered. The transmission order is synchronous, meaning that two-way interaction occurs in sequenced steps—no packet is sent unless requested, whereupon the packet is transmitted only to the requesting device, which then acknowledges receipt.

But when it comes to handling individual packets, TCP is said to be **stateless,** meaning it operates in a chaotic manner that allows any packet to be delivered regardless of packet order.

UDP SOCKET FUNCTION
A UDP socket can bind a port to unicast, broadcast, or multicast address.

UDP is TCP's sister protocol, because they share the same unicast address format and the same port number format. Nevertheless, UDP uses a different type of socket for multicast messaging. Multicast messaging is ideal with UDP's "lean-and-mean" approach to communication. Basically, UDP slashes overhead by eliminating the three-way handshake. Plus, the amount of data transmitted in a UDP packet is relatively small. If the payload does not fit entirely in the frame's user data field, it will be dropped. The UDP frame only contains source and destination port numbers, packet length number, and the data (Figure 9–9).

Because UDP does not divide a large message into smaller packets, it does not hassle with packet sequencing. No time is wasted numbering and counting packets. Instead, UDP provides a limited queue to process packets in the order received.

Without packet sequencing to worry about, UDP is not concerned about errors in packet delivery. Each UDP packet must find its own way to the destination. No retransmissions are attempted if a packet is lost or garbled.

Obviously these procedures—or lack of procedures—improve performance by reducing bandwidth requirements and speeding up program execution. Consequently, UDP is ideal for time-critical data transmission. It can handle an update for an I/O node, a PLC-to-PLC register exchange, or returning a value from a device several times faster than TCP. But since flow and error control is lacking, UDP in itself can not ensure deterministic data delivery.

To compensate, reliability services must be included in application-layer or session-layer protocols. Several iE protocols that employ UDP—such as Ethernet/IP—employ proprietary, upper layer procedures to establish and manage the connection, provide flow and error control, and guarantee delivery.

UDP header:

0	Protocol id	Length
Source address		
Destination address		
Data . . .		

TCP header:

Source port	Destination port	
Sequence number		
Acknowledgment number		
Length	Reserved	Flags		Window
Check sum		Urgent pointer
Options		Padding
Data . . .		

FIGURE 9–9   Comparing UDP and TCP headers

SOCKET PROGRAMMING CONSIDERATIONS

The choice of writing stream or datagram sockets depends on the type of transmission required. An application that opens a TCP socket for unicast messaging will be using a slower, one-to-one connection to communicate, which is appropriate for file transfers (FTP) of recipes to a PLC or for slow I/O scans in 10 ms to 50 ms. An application that opens a UDP socket for multicast messaging will be using a faster, connectionless method to communicate, which is appropriate for fast I/O scans under 10 ms—presuming there is no traffic within the collision domain that could cause dropped packets.

A nonindustrial example of multicasting is a stock ticker program. These applications transmit a single stock quotation that can be routed and displayed on many desktop PCs. In factory automation and control, UDP multicast messaging makes it possible for a single transmission from a PLC to update all the registers in every device on a subnet simultaneously.

Two related terms are used to describe this messaging style. Multicast messaging is said to be **asynchronous,** meaning that the client and server do not interact within a sequenced interval—packets are sent out like e-mail to be received at any time. Because a multicast packet can have an arbitrary number of receivers, it is also said to be **loosely coupled**—there is no dedicated, one-to-one connection between the sender and receiver.

SOCKET IMPLEMENTATION

Sockets are used in a wide range of basic client/server applications, because the technology is well known, efficient, and (when properly debugged) a reliable solution for interprocess communication. But as an early form of IPC, TCP and UDP sockets have several limitations, depending on the implementation.

TCP socket limitations

- *Relatively big and slow.* In certain circumstances, TCP's three-way handshaking and packet sequencing are **overhead,** meaning superfluous to the task. As for packet size, the TCP frame contains fields that bloat the packet. In the example shown in Chapter 1, transmitting two bytes for the value 98 takes over 50 bytes of code in the TCP frame.
- *Bandwidth hogging.* TCP's three-way handshake occurs quickly, but it does take time and uses network resources. In circumstances where an error occurs—for example, when a packet is lost—TCP can tie up a connection for minutes as it generates requests to retrieve the missing packet. TCP can further complicate problems by triggering slow-start algorithms or retransmission timeout values, and thus dramatically choke network throughput.
- *Blocking.* In circumstances where TCP's request for data has not been satisfied, it may hang (pause) an application program until it receives the reply. This behavior is known as **blocking.** There are some partial solutions. For example, to prevent TCP from hanging an application, the program can employ a **nonblocking** socket. As mentioned earlier, a socket is the communication end point created when an application opens a communication session to send and receive data from the network. A nonblocking socket uses a timer that limits the period an application waits for a reply. Or a nonblocking socket can be written that lets the application program keep running without a response from the remote program.

UDP socket limitations

- *Unreliable.* While avoiding the overhead of the TCP frame and connection-based processes, using UDP for a connectionless socket requires an application-layer mechanism to handle error correction, delivery, and other reliability issues.

General limitations of sockets

- *Security.* Dynamic port assignments will prohibit use through a firewall.
- *Not object-oriented.* Of limited use with objects, agents and other technologies developed for distributed-computing applications.

SUMMARY OF SOCKETS
Sockets are an efficient, albeit limited, IPC tool.

Features of TCP sockets

- One-to-one connection between client and server
- Unicast messaging
- 3-way handshake establishes connection with server
- Server must maintain listening port
- Server creates new socket for client
- Synchronous, blocking behavior waits for connection or data response

Features of UDP sockets
- No handshaking
- No connection between client and server
- Multicast messaging
- Self-directed packet includes IP address and destination port
- Unreliable delivery
- Asynchronous, nonblocking behavior does not wait for data delivery

Remote Procedure Call (RPC): Used in PROFInet

An RPC communication transaction occurs as a single request and a single reply in a unicast transmission.

RPC FUNCTION
As the name implies, RPCs allow a local application program to run a procedure in a program on a remote device. The procedure's parameters are prepared for serial transmission on the local host in a process known as **marshalling.** Then the code is unmarshalled on the remote system, the procedure is performed, and the result is returned.

The RPC protocol operates over TCP or UDP, because it is not concerned with message passing but only the specification and interpretation of messages involved in the remote procedure call. The RPC code itself resides in the TCP or UDP data field.

RPC PROGRAMMING CONSIDERATIONS
An RPC is generally synchronous in operation, because it blocks or halts the application program while waiting for a reply, a characteristic of a synchronous, **tightly coupled** IPC (Figure 9–10). For applications where blocking is not acceptable, an RPC can be written using lightweight processes (known as **threads**) to provide an independent execution path to keep the application running.

RPC has no mechanisms to ensure reliable transmission, just marshalling and send/receive mechanisms, and is therefore said to operate at a low level in the protocol stack. Consequently with RPC, the quality of service must be 100% reliability. That is no problem if TCP/IP is employed. With UDP/IP, however, the application layer must include protocols for retransmission and time-outs to prevent the program from hanging if a reply never comes. RPC services are separated from specific port numbers through a port mapping call to port 111, which provides port numbers for various services.

RPC IMPLEMENTATION
RPCs work well in small, point-to-point applications. An RPC's low overhead and small header provide efficient transmission (low latency). RPCs do not work well with firewalls, because too many ports must be opened to accommodate RPC's dynamic and random port

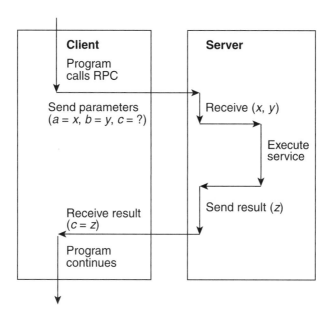

FIGURE 9–10 Synchronous RPC call

assignments. While not suitable in itself for applications involving distributed objects or object-oriented programming, an RPC can be effectively combined with other IPC mechanisms, such as DCOM and publish/subscribe. An RPC can also be used over the Internet in conjunction with XML and HTTP.

Summary of RPCs

RPC is an efficient request-reply protocol that uses a unicast transmission to carry data parameters, not large-sized messages.

Features of RPCs
- Efficient, low overhead
- Employ UDP for connectionless, asynchronous communications and TCP for connected, synchronous communications.
- The sending (client) program is typically blocked to await a reply.
- Application-layer protocols are required to handle time-outs and retries.
- Combined with other IPC mechanisms, such as DCOM and publish/subscribe

Object Request Brokers/Components: Used in PROFInet and Ethernet/IP

Object Request Brokers (ORBs) are used to handle multiple, point-to-point transactions.

THREE TYPES: DCOM, CORBA, JAVA RMI

There are three competing ORB standards:

1. *DCOM (Distributed Component Object Module).* An extension of Microsoft's popular Component Object Module (COM) specification that enables COM components to interoperate across a network.
2. *CORBA (Component Object Request Broker Architecture).* A standard backed by about 70 companies that formed the Object Management Group (OMG).
3. *Java Remote Method Invocation (RMI).* A standard backed by Sun Microsystems that enables network communication between Java programs.

COM and DCOM are popular in iE protocols due to the prevalence of Windows-based applications and control systems in the factory. All ORB technology is complex. DCOM is doubly complex due to the continuing evolution of Microsoft object-oriented technologies.

(These technologies now encompass COM, DCOM, OLE, ActiveX, and SOAP. It may be helpful to know that COM is not a programming language, but a model for defining and manipulating objects. DCOM was developed to interact with COM objects on a remote machine and was previously called "Network OLE." OLE [Object Linking and Embedding] uses object-oriented programming to embed applications within other application forms—such as spreadsheets within documents. For distributed computing, those self-contained applications evolved into ActiveX controls, which are component programs that can run in a distributed application over a network, analogous to a Java applet. SOAP, which is discussed later, was developed to pass method calls over the Internet.)

DCOM FUNCTION

DCOM is synchronous and operates over a point-to-point connection. It is classified as a high-level network protocol, because DCOM is built on top of a RPC to function as an object remote procedure call (ORPC) that transmits a method call on components residing in remote devices (Figure 9–11).

DCOM PROGRAMMING CONSIDERATIONS

DCOM is written to merge with the RPC header and data fields. It does not provide reliability mechanisms on its own, but can be used in conjunction with TCP/IP and error-handling protocols.

DCOM IMPLEMENTATION

DCOM is used in small- and large-scale applications, but requires modification to work over the Internet. DCOM dynamically assigns port numbers; consequently, a range of ports must be opened in the firewall, plus port 135, which is used for RPC end-point mapping.

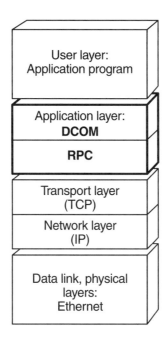

FIGURE 9–11   DCOM in the application layer

SUMMARY OF DCOM
DCOM makes it possible to invoke method calls on object-oriented programs running on remote devices.

Features of DCOM
- Enables multiple, point-to-point transactions
- Built on top of RPC to send a remote procedure call to objects
- Essential for development of object-oriented, distributed application programs on a PC platform
- Can use TCP or other application-layer mechanisms for reliability
- High overhead and firewall issues

Message Passing: Used in Foundation Fieldbus HSE, iDA, Ethernet/IP, MMS TCP/IP, ADS-net

Unlike RPC and ORB protocols, message-oriented middleware (MOM) uses multicast transmission to communicate asynchronously from one point to many points. The MOM family is comprised of messaging-passing protocols, which are used in **push** programs to send data to designated devices without a specific request, and messaging-queuing protocols, which require data to be retrieved by the application. The pub/sub messaging-passing protocol is used by many legacy fieldbuses and in the application layer of several iE protocols (Figure 9-12).

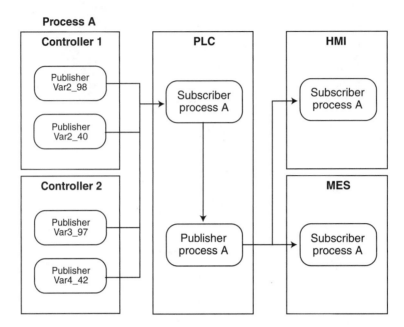

FIGURE 9–12   Publish/Subscribe messaging

PUB/SUB FUNCTION

Pub/sub overcomes the limitations of sockets and RPCs, because it does not need specific IP source and destination addresses to establish a one-to-one connection. An analogy illustrates the difference between RPC and messaging-passing procedures: an RPC communication is a private conversation between two friends; pub/sub communication is conducted in a party atmosphere—any number of interested guests can participate in the conversation at any time.

Because message-passing transmission pushes a single message to a group of devices rather than to a single host, pub/sub runs on top of UDP to take advantage of its ability to multicast to a number of devices on a subnet. Consequently, the connection with those devices is loosely coupled. That can be an advantage in industrial control, because loose coupling allows a single transmission to be picked up by many devices within a short period, avoiding the network congestion that comes from updating and polling tightly coupled devices one at a time.

PUB/SUB PROGRAMMING CONSIDERATIONS

Pub/sub messaging is not based on a one-to-one connection using the IP addresses of two devices. Instead, pub/sub requires that devices register their interest (subscribe) to a set of topics. The topics can be subject tags that are determined by the programmer —such as "reactor temperature"— to relate the value 98 to a PLC table or SCADA database.

With pub/sub, the programmer avoids the tedious assignment of device addresses typical with socket-based applications. The middleware maintains the topic list in a database, changes IP address as devices subscribe and unsubscribe to topics, and routes the message to all subscribing devices.

A device that transmits values with a topic identifier is known as an information **producer.** Transmissions can be based on events or a fixed schedule. Devices that have indicated their interest to receive messages on a topic are information **consumers.** The topic tag is retained in the device's local memory or the memory of a server or router. Subscribing nodes will notice the identifier, pull the packet off the wire, and consume its contents.

PUB/SUB IMPLEMENTATION

Pub/sub is appropriate in large-scale, distributed applications. Message-passing middleware is not recommended in situations where processes are disconnected for long periods.

Publish-subscribe is typically implemented with a set of agents that maintain the information in a real-time database. The application-layer protocol establishes a connection with one of the agents and sends the message to it. The agent then routes the message.

In place of TCP, pub/sub middleware handles reliability issues, such as error correcting, data flow rate, and dynamically routing messages along the most efficient path.

There are two architectures used with pub/sub. A hub-and-spoke (star) architecture is employed in business applications. Application clients are connected only to a server for message distribution. If any connection fails, the message server stores messages until it is restored. Where a firewall is involved, ports must be opened to allow message-passing for the server.

A bus architecture is used in real-time industrial networking applications. In such cases, there is no centralized message server to coordinate the distribution of messages. Instead, clients perform message-server functions. All clients are registered in an IP multicast group, which functions as the network bus. Message routing is handled by this layer, while the client application layer is responsible for reliability. Because all the necessary client/server messaging services are retained in the local devices' memory, pub/sub performance on a bus architecture is very fast. Firewall security is a drawback, however, because administrators must open ports to allow multicast message traffic to pass. For that reason, this model is not recommended for Internet deployments.

PUB/SUB SUMMARY

Publish/subscribe is message-passing middleware that employs multicast transmission for asynchronous communication from one point to many points.

Features of Pub/Sub
- Substitutes laborious assignment of IP addresses with plain language topics
- Suitable for objects and agents
- Reduces network traffic by avoiding multiple transmissions of identical messages
- Transmissions to be synchronized by schedule or by event
- Needs a gateway or Web portal for Internet access

SUMMARY

- Ethernet and TCP/IP technologies work fine—it is the different application layer protocols created by vendors that cause chaos.
- The application layer provides the specific communication services required when transmitting data between different devices.
- The requirements of application programs directly impact the design of the protocol stack, especially the application layer.
- In the point-to-point model, information is exchanged between one sending application and one receiver.
- The publish/subscribe (pub/sub) model allows a single sending application to transmit one message to several receiving applications.
- Application programs can be written using one of four programming approaches, each of which is predicated on a specific communication model.
- For network communications, several kinds of technologies were developed—based on sockets, remote procedure calls (RPC), object request brokers (ORBs), and pub/sub—which are collectively known as middleware.
- Middleware is employed in interprocess communication (IPC) to enable processes in different programs to run concurrently and to exchange data with each other.
- Sockets are the primary tool used for interprocess communication between devices in a client/server relationship.
- TCP is used for unicast messaging; UDP for multicast messaging.
- RPC is an efficient request-reply protocol that uses a unicast transmission to carry data parameters, not large-sized messages.
- DCOM makes it possible to invoke method calls on object-oriented programs running on remote devices.
- Publish/subscribe is message-passing middleware that employs multicast transmission for asynchronous communication from one to many points.

? REVIEW QUESTIONS

1. What are the three addressing methods supported by IPv4?
2. Why are the presentation and session layers not included in the iE stack?
3. What is interprocess communication?
4. There are four approaches to programming—what IPC techniques are generally associated with each one?
5. What is the most common type of IPC programming?
6. What is a socket?
7. What type of IP address is used by a multicast socket?
8. How is the Ethernet header modified to indicate a multicast packet?
9. Why is TCP stateless?

10. Why is UDP ideal for real-time communication?
11. Define **blocking.**
12. What is the function of an RPC?
13. Does an RPC use blocking?
14. Is an RPC reliable?
15. What message size is suitable for RPC?
16. How does DCOM use RPC?
17. Is DCOM easy to use over the Internet?
18. Give an example of a **push** program.
19. Why does publish/subscribe use UDP and multicasting?

Nine iE Protocols: How They Stack Up

10

OBJECTIVES

After reading this chapter, you will be able to:

- Use a business criteria to measure nine iE protocols
- Understand the pros and cons of five TCP/IP-based iE protocols: Modbus/TCP, PROFInet, INTERBUS, MMS TCP/IP, and .NET for Manufacturing
- Understand the pros and cons of four iE protocols that use UDP/IP in whole or in part: Foundation Fieldbus HSE, Ethernet/IP, iDA, and ADS-net
- Summarize iE integration and application strategies

The previous chapter looked at the core technologies used in the network, transport, and application layer for interprocess communication. Now it is time to see how they work in specific iE protocol stacks.

Note: The information that follows is derived from publicly available resources that are subject to change. In almost every case, the protocol specification is available through the referenced Web site. All decisions should be based on the most current standards in conjunction with the advice of qualified professionals. Neither the authors nor the publisher warrant the accuracy of the following descriptions—your own research is the best guide.

REVIEW CRITERIA

There are several ways to evaluate iE protocols. One way is to employ the same criteria used to measure a fieldbus's suitability for:

- Discrete or process control
- I/O updates or recipe transfers

- Real-time response or synchronized messaging
- Scalability and flexibility
- Embedded communications
- Distributed applications
- Intrinsic safety and security
- Cabling distance
- EMI protection

For each of these concerns, iE networking is as fit as any fieldbus at the control level and often at the I/O level—thanks to the evolution of cabling, connectors, switches, and gateways, full-duplex communication that can handle the demanding industrial environment. Ethernet interfaces are commonplace in controllers, I/O modules, and other devices.

Any misgivings about adopting iE is usually not due to technical, economic, or performance issues, but to broad business considerations—such as ROI, security risks, and the potential of disruption and downtime in the plant. Consequently, this chapter also takes a broader view by examining an iE protocol's historical roots, present capabilities, and future potential. Four basic questions are asked:

1. What encapsulation technique is used?
   The answer will show how closely an iE protocol is tied to its predecessor fieldbus protocols.
2. What is the network/transport mechanism?
   The answer will determine how the protocol balances the trade-off between speed and reliability.
3. What messaging model is used?
   The answer will indicate the protocol's flexibility and scalability as the business grows.
4. Does it support profiles and objects?
   The answer will determine the protocol's suitability for developing distributed applications and management tools that make life easier for control engineers, programmers, system integrators, and users.

Encapsulation Technique

Encapsulation is a quick and easy way to get a legacy fieldbus protocol to run on Ethernet. The developer only needs to engineer a way for the legacy protocol to be nested within the TCP or UDP frame. As was previously shown, the fieldbus equivalent of a datagram, called a **telegram,** can simply be embedded into the user data field of the IP datagram.

While this approach makes it possible to get an iE version of a legacy protocol to market quickly, encapsulation requires that the user employ applications that can extract and process user data in the format defined by the original protocol.

Network/Transport Mechanisms

In the trade-off between speed and reliability, TCP/IP sacrifices latency and overhead for establishing a reliable connection and handling errors. In some situations, the trade-off is a bad deal: with TCP/IP, minutes can go by before the problem becomes evident. Such delays are not acceptable for time-critical applications. In these cases, UDP/IP must be used in conjunction with error correction mechanisms in the application layer to combine reliability and real-time communication.

Messaging Model

Devices can relate to each other as client/server, master/slave, peer-to-peer, or producer/consumer. The model that is adopted is usually the offspring of the legacy protocol: for example, master/slave communication has migrated to source/destination messaging based on a client/server model. As previously discussed, source/destination messaging sends a separate message to each device even if the contents are identical. Publish/subscribe (pub/sub) messaging is a more recent development based on a publisher/consumer model used in several fieldbus protocols and adopted for many iE protocols. Pub/sub allows a device to send a single message that can be read by multiple devices simultaneously.

Profile/Object Support

A device profile is an abstraction used to determine what kind of input signals and instruction will produce the desired output signals and results. A device profile can be programmed in several forms. For example, I/O interactions can be written as function blocks, which are standardized, diagrammatic representations of logic used to process variables and to build device-management tools.

Accordingly, profile/objects are not in the iE protocol itself, but reside in the user layer of the application program or in the firmware on board-intelligent devices. These programs manage devices using function blocks for communication, alarms, and process variables to operate the device—plus function blocks for the identification number, input channels, alarm status, and address to define the device.

In contrast to function block code, object-oriented programming creates a model of the function profile, based on a collection of objects and interactions among the objects. Each device type correlates with a profile, which is associated with the appropriate class of objects. This approach makes it possible for different devices of the same type to be interchangeable, because interaction occurs between common objects, not between the hardware's particular data type.

Device profiles can be implemented very simply by providing the device's identity and the relevant device parameters. A full object model, however, also provides a common definition of behaviors to simplify programming. For example, a full object model for a motor starter can include an identity object, an overload object, and a discrete output object. Since the overload object includes configuration options that are common to the overload-protection function of several different motors and drives, it can be simply reused. The programmer does not hassle with customization and the user is not bothered by inconsistencies.

In distributed control, the issue is how to get one device to activate code that resides on another device to perform the required action. Consequently, an iE protocol that is used in distributed, object-oriented applications must support remote procedure calls on objects.

TCP-BASED iE PROTOCOLS

Thanks to the connection-oriented reliability of TCP/IP, it is employed as the core technology in five iE protocols: Modbus/TCP, PROFInet, INTERBUS, MMS TCP/IP, and .NET for Manufacturing (Figure 10–1).

Modbus/TCP *(Backed by Schneider Electric, documentation is available through www.modbus.org)*

Modbus/TCP was the first open iE protocol on the market. Based on the popular Modbus protocol used in industrial automation, it was written by Andy Swales of Schneider Electric and was released by Modicon/Groupe Schneider in 1999 as Open Modbus/TCP. The simplicity of Modbus/TCP makes it a strong competitor against more complex and more capable iE protocols.

ENCAPSULATION TECHNIQUE

In developing Modbus/TCP, Modicon took the Modbus frame used in RS-232/RS-485 serial communications and lifted it on top of the TCP/IP platform practically "as is" (Figure 10–2). The telegram is encapsulated within the TCP/IP datagram's user data field.

There are a few minor differences between the telegram and datagram versions. In the legacy Modbus telegram, the initial field contains the slave device's address, with two ASCII characters used in Modbus ASCII addressing or eight bits used in Modbus RTU (remote terminal unit) addressing. For Modbus/TCP, this address field is replaced by a one-byte unit identifier to create an IP address and unit ID combination that uniquely identifies the device on the network. The function field contains code telling the device what action to perform. The data field contains additional information—such as the master telling the number of actions to perform or the slave providing the requested data. Also, error codes or no code may be in this field, depending on the type of messaging. The Modbus 16-bit checksum is replaced by TCP's 32-bit CRC.

The straight-forward approach of putting the Modbus frame inside a TCP/IP wrapper is very appealing to developers. Any competent programmer can implement Modbus/TCP using Modbus serial source code and appropriate TCP socket software.

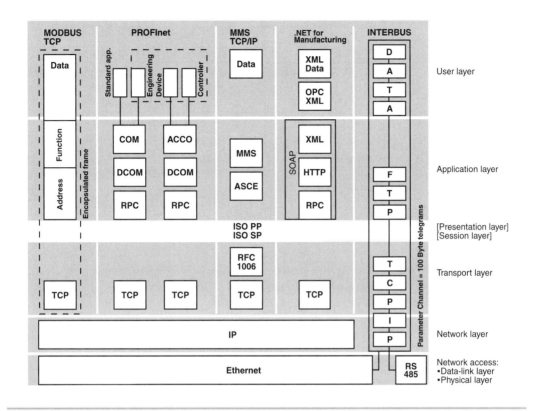

FIGURE 10–1 Comparison of TCP/IP-based iE protocols

ACCO: Active Control Connection Object	ISO PP: International Standards	OPC: OLE (Object Linking and Embedding)
ASCE: Association Control Service Element	Organization Presentation Protocol	for Process Control
COM: Component Object Module	ISO SP: International Standards	RPC: Remote Procedure Call
DCOM: Distributed Component Object Module	Organization Session Protocol	TCP: Transmission Control Protocol
FTP: File Transfer Protocol	MMS: Manufacturing Message Service	UDP: User Datagram Protocol
		XML: Extensible Markup Language

NETWORK/TRANSPORT MECHANISM

Modbus/TCP fully exploits TCP/IP's connection services to ensure every query receives a reply. In fact, it nicely complements the point-to-point connection of legacy Modbus. In the master/slave relationship, a single application controls communications with all devices. Where the Modbus master controls network access to avoid collisions that would otherwise hamper message delivery, now Modbus/TCP uses Ethernet's collision detection and avoidance mechanism. Where the Modbus master controls error checking, now Modbus/TCP uses TCP/IP.

Modbus's master/slave architecture is adapted for TCP/IP's client/server model with a few modifications. At the application layer, a fragmentation protocol is added to maintain

Modbus frame

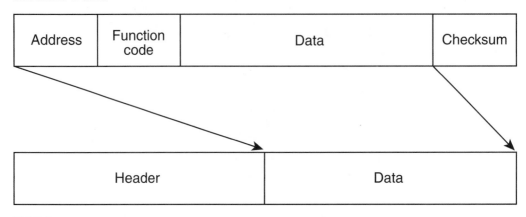

TCP frame

FIGURE 10–2   Modbus frame encapsulated within the TCP frame

compatibility with legacy Modbus applications. Modus RTU-to-TCP gateways can be used to allow Modbus/TCP applications to access binary data from Modbus RTU serial devices. A Modbus HMI or PC master is converted into a client by using an Ethernet interface and a Modbus/TCP protocol stack. Similarly all Modbus slaves can be provided with device server support to convert them into Modbus/TCP servers.

Since Modbus/TCP makes no distinction between master and slave, any node with a TCP port can access any similarly equipped node. Some vendors implement peer-to-peer communication for Modus I/O. Tens of thousands of nodes can be added using Ethernet switches to create small domains to avoid collisions that would otherwise hamper deterministic performance. Modbus/TCP packets can be routed through firewalls by opening port 502.

MESSAGING MODEL

In the legacy Modbus communication model, there is no uncertainty: The master requests; the slave replies. The data flow is one direction only. This model has been adopted for the iE version, which takes advantage of TCP/IP for reliable point-to-point connections, unicast transmission, and source/destination messaging.

Modbus/TCP has been criticized for depending on TCP's tedious bookkeeping habits when initiating a communications session and handling packets. It is also criticized for depending on its static, unicast, source/destination messaging.

As discussed in the section on message-oriented middleware, message-passing protocols rely on UDP for multicast transmission. In this regard, Modbus/TCP can not compete because its underlying transport mechanism—TCP—can not do multicasting.

The critics point out three problems with source-destination messaging:

1. It is a network hog.
   In circumstances that require sending multiple copies of identical messages, Modbus/TCP must transmit the same contents to each device one by one, which needlessly hogs network resources.
2. It makes synchronized control difficult.
   Because each message is transmitted sequentially in time, devices down the line get the data later, which makes it difficult to provide the simultaneous time stamp necessary to synchronize control.
3. It does not prioritize.
   There is no distinction between large instruction-filled messages and small real-time data transmission. In circumstances where small packets containing time-critical I/O are queued up with large messages, the I/O must literally take a number and wait its turn in the communication sequence, which makes real-time control difficult. In these applications, separate networks may be required: one for messaging and one for real-time data.

In response to this criticism, Andy Swales defends unicast, source/destination messaging against UDP multicasting, noting that:

[M]ulticasting is a disaster on large layer 2 switched networks. A layer 2 switch has no means of determining whether an incoming multicast message has any interested receivers on a particular outgoing port. Therefore, it is forced to treat the multicast just like a broadcast, causing the message to be copied on all ports "just in case."

When you [have] 100 users, each subscribing to 5 or 10 of these incoming message streams, the network load soon [adds] up. And unfortunately, EVERY message got copied to EVERY station on the switched Ethernet, since the messages were multicast.

The producer-consumer proponents, as a defense, suggest that large networks always should be subdivided using layer 3 routers, which will make informed decisions about whether to copy multicast messages using the protocol called Internet group management protocol (IGMP). And if you require access to multiple IP subnets at a physical location, you can always use virtual local area network (VLAN) techniques, which are supported by the router and switch manufacturers.

Unfortunately, the cure is worse than the disease. Using IP subnets to partition the network makes mobility impossible. And requiring administration of VLAN tables and routing tables to make it work leads to excessive bureaucracy and administration.

We report, you decide.

PROFILE/OBJECT SUPPORT

According to control developer Lantronics, Modbus "uses a simple 16-bit number to represent a reference number and count to specify a requested data transfer to or from a device."

Consequently, Modbus/TCP implements device profiles very simply by providing the device's identity—such as a tag identifying motor starter—and the relevant device parameters.

This technique makes it difficult to correlate any changes between device tags and corresponding information in HMIs or device memory maps.

Modbus/TCP has been criticized for not fully supporting objects. According to SEMA-TECH, however, Modbus/TCP can be implemented with object properties allowing the exchange of data items using class, instance, and attribute number. The Modbus/TCP object model protocol is based on the SEMI SensorBus system.

While not a full object model, this approach allows the programmer to make changes on the basis of a hierarchical tag name rather than laboriously looking up a device's register number and altering it in the client program.

PROFInet *(Backed by Siemens, documentation available through www.profibus.com)*

ENCAPSULATION TECHNIQUE

While PROFInet is the outgrowth of Profibus International's legacy PROFIBUS protocol, the organization decided not to encapsulate PROFIBUS in a TCP/IP frame. Instead, the PROFInet Users Organization developed an abstract interface in the application layer similar in design to OPC (OLE for process control). This high-level interface enables PROFIBUS and other devices that implement the standard to interact independently of underlying protocols to promote interoperability between different devices. In fact, PROFInet can be implemented on any real-time operating system for PLCs that can handle a TCP/IP stack and common DCOM and RPC protocols. With engineering system application packages, however, PROFInet should be implemented on a Microsoft platform using COM, OLE, and IDL (interface definition language). A gateway must be used to communicate with PROFIBUS devices.

NETWORK/TRANSPORT MECHANISM

PROFInet uses TCP/IP to maintain a reliable connection. For real-time communication, UDP/IP is still an option, because PROFInet devices negotiate the optimum protocol—TCP/IP or UDP/IP—while establishing the relationship between interfaces. To maintain the link with the interface, a special technology is employed—the active control connection object (ACCO). ACCO ensures reliable data transfer by establishing the relationships between the interface, by exchanging data, and by monitoring the interface partners. ACCO also handles errors by transmitting a quality code and time stamp, re-establishing lost connections, and providing diagnostic tools. To communicate over Ethernet with PROFIBUS devices that do not use the interface and with third-party protocols, a proxy is required.

MESSAGING MODEL

PROFInet does not use the pub/sub messaging model, but rather employs DCOM to handle interactions between objects and methods. Object remote procedure calls (ORPC) are used for event mechanisms and method calls to remote devices. A **runtime model** is installed on each PROFInet device, which is an executable version of code that provides common points of interaction between components. In the PROFInet model, a component combines physical (conveyor), electrical (drive), and programmatic (control program) elements into an entity that can be connected with other components to establish a communication path within the manufacturing cell and between the MES and ERP systems. The use of DCOM makes it difficult to route PROFInet packets through a firewall. Implementing gateways with other busses is complex.

PROFILE/OBJECT SUPPORT
PROFInet seeks to exploit the full potential of object-oriented programming to open up PROFIBUS to communicate at every level of the organization. To accomplish this, the application layer can handle four different COM interfaces: a mandatory runtime interface implemented on all devices, a device-specific interface implemented in the firmware, an application-specific interface using programming tools that are specific to the destination system, and optional interfaces. In operation, PROFInet correlates or **maps** runtime objects on devices with engineering system objects on HMIs and workstations, like a Word document uses OLE to update an embedded Excel spreadsheet. Different vendors can use XML to write descriptions of device functions into component objects. An engineering tool's graphical user interface lets programming engineers open the objects and create communication links between components by simply drawing lines.

A user friendly development environment is particularly helpful given the complexities of interacting with the PROFIBUS protocol structure, which includes PROFIBUS FMS (fieldbus message specification). This universal protocol, which is specified in DIN 19245, provides cell-level communication between PLCs and is based on an object-oriented client/server model that was derived from the MMS standard described below.

INTERBUS *(Backed by Phoenix Contact, documentation available through www.ibsclub.com)*

ENCAPSULATION TECHNIQUE
The INTERBUS protocol, developed by INTERBUS club members, takes a unique approach to encapsulation. Instead of embedding an INTERBUS telegram inside the TCP/IP data field, this technique divides up the datagram for transmission over a 100-byte INTERBUS frame field known as a *parameter channel*.

NETWORK/TRANSPORT MECHANISMS
To transfer a single TCP/IP frame within its 100-byte parameter channel field, INTERBUS must use a number of transmission cycles. TCP/IP transmissions, therefore, do not occur at Internet speed, rather at the speed of INTERBUS—550 kb/s at the standard rate, which is sufficient for real-time communication for INTERBUS serial communication. To handle larger message sizes and the overhead of TCP/IP headers, a new, higher rate of 2 Mb/s is available.

INTERBUS uses multiplexing to accommodate a hybrid protocol structure that can transmit INTERBUS process data, parameter data, and TCP/IP file transfers over the same RS-485 bus. To encode the TCP/IP telegrams into standard packets for transmission over Ethernet is not very time consuming. Nevertheless, an Ethernet/INTERBUS gateway is required to connect an INTERBUS system to Ethernet and to exchange INTERBUS data with a host computer on an Ethernet LAN, WAN, or Internet. Transmitting packets through a firewall is a matter of opening ports used by the gateway.

MESSAGING MODEL
Strictly speaking, INTERBUS uses TCP/IP in conjunction with FTP to transfer messages to PLCs and controllers. Therefore from the viewpoint of the iE protocol stack, messaging is handled by FTP in the application layer, not any IPC mechanism. Access through a firewall is simply a matter of opening up FTP port 21.

PROFILE/OBJECT SUPPORT
The INTERBUS standard includes a wide range of well-established device profiles. Device manufacturers are responsible for determining object values in an object dictionary.

MMS-TCP *(Based on ISO/IEC 9506 standard, documentation available through www.nettedautomation.com)*

ENCAPSULATION TECHNIQUE
The Manufacturing Message Specification (MMS) is an outgrowth of the General Motors-backed Manufacturing Application Protocols (MAP) initiative begun in the 1980s to define communications for a plethora of proprietary protocols for numerically controlled equipment, PLCs, and robots. The MMS documentation was initially released in December 1988, but not as an Ethernet-based protocol. The core protocol document was published in 1990 as ISO/IEC 9506-2. A second version was published in 1999. Not limited to manufacturing applications, MMS is widely employed in the utility communication architecture (UCA) used by the electric power industry.

Developers have only recently implemented MMS on top of an iE protocol stack. In this case, the MMS structure is encapsulated at the application layer, except where it is redesigned for object-oriented messaging.

NETWORK/TRANSPORT MECHANISMS
Because it is an abstract messaging specification that just provides syntax rules, objects, and services for accessing and exchanging information, MMS must be implemented on top of a reliable network and transport mechanisms. MMS doesn't depend on a specific protocol stack, so Ethernet-based versions have been developing using TCP/IP or ISO/OSI connection and reliability mechanisms. Firewall issues depend on the implementation.

MESSAGING MODEL

There are three ways MMS can be implemented over TCP/IP:

1. Use ISO layers on top of TCP/IP to maintain full conformance with the MMS standard.
2. Use RPC-based middleware to maintain the MMS standard's packet structure, while losing OSI session, presentation, and service mechanisms.
3. Use CORBA to implement a version of MMS redesigned for an object-oriented environment.

PROFILE/OBJECT SUPPORT

Like PROFInet, MMS uses data modeling to create a message standard for an abstract device, which can then be customized for specific device types. Services include parameter access, event management, and file transfer. Device type descriptions exist for robots, numerical control machines, PLCs, and process control and supervision, including a function block model.

.NET for Manufacturing *(Backed by Microsoft, documentation available through www.microsoft.com/technet/itsolutions)*

ENCAPSULATION TECHNIQUE

.NET for Manufacturing is a Microsoft platform that uses simple object access protocol (SOAP) for loosely coupled, asynchronous communication with applications across the enterprise. In contrast, Windows DNA technology (COM, DCOM, Visual Basic) is used for tightly coupled distributed applications on the factory network. SOAP overcomes DCOM's inability to work across the Internet through firewalls or through network address translation (NAT) software, plus provides the convenience of XML for data tagging and device description. The initial SOAP standard was published in 1999 by Microsoft, Don Box, Dave Winer, and others. SOAP is now a project of the W3C's XML Protocol Working Group.

Because SOAP is designed as an independent platform, every function call that can run code on a remote host can be encapsulated in a SOAP envelope. This includes RPC and COM/DCOM calls.

NETWORK/TRANSPORT MECHANISM

SOAP messages are transmitted as HTTP pages running on top of TCP/IP, the same protocol stack used in Internet communication. A protocol is being developed to allow for SOAP messages to be transmitted directly over TCP or UDP without HTTP. Access through port 80 on a firewall is usually not an issue.

MESSAGING MODEL

To send messages through port 80, SOAP uses HTTP's services for HTML documents. But instead of HTML, SOAP uses short-lived documents tagged with XML code to trigger operations and responses on remote hosts. RPC request and response messages are encoded as XML structures.

OBJECTS/PROFILES SUPPORT

SOAP is object neutral and can interact with COM, CORBA, or Java RMI components. While SOAP has a defined Web Services description language, a device description language suitable for factory control and automation must be defined by the developer—a likely source of conflict between applications.

UDP-BASED iE PROTOCOLS

Either in whole or in part, UDP is the transport-layer protocol used in four iE protocols: Foundation Fieldbus HSE, Ethernet/IP, iDA, and ADS-net (Figure 10–3).

Ethernet/IP *(Backed by Rockwell Automation/Allen-Bradley, documentation available through www.odva.org)*

ENCAPSULATION TECHNIQUE

Ethernet/IP (here IP stands for *industrial protocol*) was developed by the Allen-Bradley Company (part of Rockwell Automation) and is now maintained by the Open DeviceNet Vendor Association (ODVA). It is based on ControlNet and DeviceNet fieldbus protocols, which are popular in discrete manufacturing.

Ethernet/IP embeds its legacy fieldbus telegram either in a TCP or UDP frame depending on the service required. DeviceNet and ControlNet protocols are based on the Control and Information Protocol (CIP).

NETWORK/TRANSMISSION MECHANISMS

In fieldbus applications, CIP is a versatile protocol that can handle large messages for instructions, small I/O data, as well as interprocess communication between objects. Therefore, encapsulated CIP telegrams include a header to indicate which quality of service is required, either TCP for bulk messages or UDP for real-time data exchange. The control portion of CIP is used for real-time data transport employing implicit messaging. The information portion of CIP is used for less-time-sensitive transport employing explicit messaging (see Figure 10–4).

Ethernet/IP is promoted for three specific capabilities:

For information.   An explicit, point-to-point connection is established to transmit data involving large-packet sizes, such as batch recipes. It is suited to operations requiring a reliable connection more than real-time performance. A short-lived connection is established between one originator and one target (destination) device. The reliability services inherent in TCP/IP protocol ensure delivery.

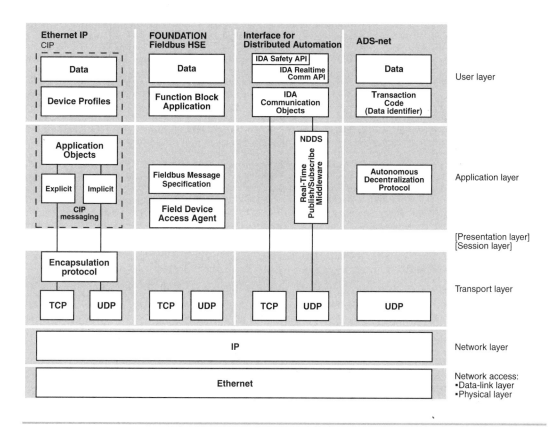

FIGURE 10–3   Comparison of UDP/IP using iE protocols

CIP: Control Information Protocol
COM: Component Object Module
DCOM: Distributed Component Object Module

UDP: User Datagram Protocol
XML: Extensible Markup Language

Transmission protocol	Connection	Ethernet/IP usage	Example
TCP	Explicit point-to-point	For information	Large data blocks, such as recipes
UDP	Implicit point-to-point or producer/consumer	For real-time I/O	Small, byte-size I/O data
UDP	Implicit producer/consumer	For real-time interlocking (cyclic), polled, and events	Exchange data between multiple devices

FIGURE 10–4   Ethernet/IP transport mechanisms

For real-time I/O.   An implicit connection using UDP/IP enables operations requiring time-critical performance, such as I/O updates, rather than bulk data transfers. Because I/O data is typically byte-sized, UDP/IP protocol is used data for high-speed delivery.

For real-time interlocking (cyclic), polled, and events.   An implicit connection using UDP/IP multicasting allows data transmission between one originator and multiple target devices. It is well suited to the cyclic data transfer necessary for synchronizing operations between one producer device and many consumer devices. High-speed UDP datagrams are used for scheduled polling of slaves by masters for cyclic "heartbeat" messaging of slave status, and for event messaging notifying change of state.

With Ethernet/IP, transmission through a firewall is an issue depending on the implementation.

MESSAGING MODEL

Allen-Bradley originally introduced products for Ethernet known as Ethernet PLC and SLC that used client server protocol (CSP) for master/slave communication. But to remain interoperable with CIP employed in its legacy ControlNet and DeviceNet protocols, Ethernet/IP adopted CIP's producer/consumer model. A single Ethernet/IP transmission can be sent to multiple devices multicast at one time using implicit messaging enabled by IP's multicast services. It is important to distinguish, however, that Ethernet/IP does not accommodate topics required for publish/subscribe messaging.

PROFILE/OBJECT SUPPORT

ControlNet, DeviceNet, and Ethernet/IP share the same profiles for devices and objects. Consequently, devices based on the CIP application object library are **plug-and-play;** that is, they automatically interoperate over the network. The object definitions are strictly defined to support real-time I/O messaging, configuration, and diagnostics. Consequently complex devices, such as robot controllers, drives, bar code scanners, and scales, can communicate over the same network without custom programming.

Foundation Fieldbus HSE *(Backed by the Fieldbus Foundation, documentation available through www.fieldbus.org for a fee)*

ENCAPSULATION TECHNIQUE

The Fieldbus Foundation was formed in 1994 by a coalition of 120 process control vendors interested in creating a single, interoperable fieldbus specification based on the international Open Systems Interconnect (OSI) standard. Foundation Fieldbus HSE is an 100-Mb/s iE protocol for control-level networks that link field devices to the legacy Foundation Fieldbus H1 protocol. Prior to HSE, H1 networks could only connect directly to controllers. HSE, however, allows H1 network segments to terminate in linking devices that use Ethernet to connect with controllers and HMI devices. Thus, islands of devices linked by H1 can talk over long distances (common in chemical refineries) with controllers linked by HSE.

The Foundation Fieldbus HSE (High Speed Ethernet) control system encapsulates protocols used by the legacy H1 protocol, which is popular in process industries. But several new specifications have been added to the HSE protocol stack in accord with the foundation's desires for the industrial application of Ethernet, namely redundancy and bridging capabilities with its legacy H1 protocol. Also, a flexible function block (FFB) wrapper has been developed so the H1 function block model can be used for discrete control and for creating gateways that allow other protocols to use the HSE backbone.

NETWORK/TRANSPORT MECHANISMS

Like Ethernet/IP, Foundation Fieldbus HSE employs TCP/IP and UDP/IP to handle bulk messages and real-time I/O, respectively.

For redundancy at the network layer level, Foundation Fieldbus HSE devices contain a HSE LAN redundancy entity (HSE LRE). The HSE LRE periodically sends diagnostic messages to update a network status table in their local memory. Consequently if a network segment goes down, it can attain the address of working switches to reroute the message along a different path. Firewall ports may be a problem depending on the implementation.

MESSAGING MODEL

The HSE protocol stack uses TCP/IP and UDP/IP to replicate the types of messaging supported with H1 legacy protocol. A client/server model is implemented over TCP/IP to handle unscheduled request/response messaging used for bulk file transfers between hosts and devices. A publish/subscribe model uses UDP for scheduled data acquisition and for synchronizing inputs and outputs during cyclic data transfer in the control loop between devices. The UDP multicast capability is also used for event notification, such as alarming and trending.

PROFILE/OBJECT SUPPORT

The Foundation Fieldbus standard offers a wide range of device descriptions that provide device drivers that host systems can use without the need for custom programming. While not implementing a full object model, the Foundation Fieldbus user layer supports standard function blocks to provide a consistent interface for inputs and outputs to make interoperability possible between devices from different vendors.

iDA *(Backed by Phoenix Contact, documentation available through www.ida-group.org)*

ENCAPSULATION TECHNIQUE

The iDA (Interface for Distributed Automation) protocol is under development with the goal of obtaining real-time performance at speeds allowing precise drive synchronization over the Ethernet. Highly synchronized networks are needed in motion control, robotic, and packaging applications.

Because the iDA standard is being newly developed, there is no prior fieldbus telegram to encapsulate.

NETWORK/TRANSPORT MECHANISMS

TCP/IP is used with standard Internet application-layer protocols, such as HTTP, SMTP, and FTP, for exchanging program parameters, diagnostics, and remote control. UDP/IP is used for real-time communication and services provided by iDA middleware.

MESSAGING MODEL

With iDA, real-time messaging is based on the pub/sub model employing UDP/IP, specifically the real-time publish-subscribe (RTPS) protocol. RTPS extends publish-subscribe to include the timing, reliability, and low overhead necessary for real-time applications. According to its developer, Real-Time Innovations, all RTPS transactions are time-stamped and controlled so "each subscriber can specify a data arrival deadline. If the data doesn't arrive in time, the application is notified."

RTPS is the wire protocol used by RTI's network data delivery service (NDDS) middleware, which includes management services and tools for the developer. Real-time services are used for:

1. Data distribution
2. On-demand data exchange
3. Remote method invocation
4. Event notification

PROFILE/OBJECT SUPPORT

Like PROFInet, iDA uses object-oriented programming to create an abstract interface so developers can build application programs for devices regardless of the vendor. Object methods can be run remotely via the IDA method server. XML is used to define style sheets and data structures, which are the building blocks for defining common objects, such as drive controllers. Each device must be named according to the IDA Structure.

All iDA field devices can serve up their own Web page for accessing and modifying XML-tagged configuration, operation, and diagnostic parameters using a Web browser.

ADS-net *(Backed by Hitachi and European user Beckhoff, documentation available through www.mstc.or.jp/jop)*

ENCAPSULATION TECHNIQUE

The Autonomous Decentralized System (ADS) network standard is a development of the Japan FA (Factory Automation) Open Systems Promotion Group of the Manufacturing Science Technology Center (JOP/MSTC). Hitachi is a prominent industrial supporter of the standard. The specification was published in September 1999 by the JOP/MSTC's Distributed Manufacturing Architecture Technical Committee. Encapsulation does not appear to be relevant. An FL (Factory Link) Net standard is also promoted by the JOP/MSTC, but is not discussed here.

NETWORK/TRANSPORT MECHANISMS

TCP/IP is used for FTP, Telnet, and SMTP. UDP/IP is used for ADS multicasting.

MESSAGING MODEL

Instead of a hub-and-spoke message-passing system, ADS-net employs a bus architecture for pub/sub transactions. All the necessary messaging services are retained in each node, which reduces the burden of node management and addressing in a decentralized automation system. A transaction code data (TC) identifier in the ADS data field takes the place of a typical pub/sub topic.

PROFILE/OBJECT SUPPORT

ADS-net supports a Module Object Model, but does not provide one. Device identity, device function, device management, and application process objects can be specified.

INTEGRATION AND APPLICATION STRATEGIES

The incompatible connectors, signals, and encoding associated with yesterday's fieldbus protocols are being superseded by today's iE protocols. It is obvious, however, that there are still many incompatibilities (Figure 10–5). Therefore, another middleware technology—called OPC DX—has been developed as a partial solution.

OPC DX *(Backed by the OPC Foundation, documentation available through www.opcfoundation.org)*

The OPC Foundation has taken the initiative to promote an interface for interfaces—the OPC Data Exchange (DX) standard for Ethernet. OPC DX takes advantage of the iE application layer being open and sufficiently defined to allow competing vendors to build a common middleware solution to make devices and applications interoperable. Essentially, this approach creates an abstract layer in harmony with the abstract interfaces provided by several iE protocols (see Figure 10–6).

According to the OPC Foundation, the OPC DX standard will provide interoperable data exchange and server-to-server communications across Ethernet networks. It is an extension of the existing OPC data access specification used to interchange HMI and controller data.

Specifically, the OPC DX approach involves developing an overarching application layer standard with which other standards can interact. At the time of this writing, however, only three iE protocols are involved: PROFINet, Foundation Fieldbus HSE, and Ethernet/IP.

Besides allowing interaction with devices on different buses, the OPC DX standard makes it possible to develop plug-and-play software components from different automation suppliers.

IE protocol	Encapsulated telegram	TCP/IP UDP/IP	Port usage	Messaging	Profile/object support	OPC DX support
Modbus/TCP	Modbus	TCP/IP	502	Source/ destination	Legacy	No
PROFInet	PROFIbus Plus	TCP/IP	Dynamic	Source/ destination	ORPC	Yes
INTERBUS	TCP/IP	TCP	Dynamic	Source/ destination	Legacy	No
MMS TCP/IP	MMS	TCP		Source/ destination	MMS	No
.NET for Manufacturing	COM	TCP/IP	80	Source/ destination	DCOM/XML	No
Ethernet/IP	CIP	TCP/IP explicit UDP/IP implicit	Dynamic	Producer/ consumer	Legacy	Yes
Foundation Fieldbus HSE	H1	UDP/TCP optimized	Dynamic	Pub/sub	Legacy plus	Yes
iDA	Not applicable	UDP/IP	Dynamic	Pub/sub	XML authoring	No
ADS-net	Not applicable	UDP/IP	Dynamic	Pub/sub	Possible	No

FIGURE 10–5   Different features in various iE protocols

Playing the Protocol Stack: Read 'em and Weep

While Ethernet has failed to deliver a universal, upper-layer iE protocol, it has given birth to a greater degree of openness so that users are much more aware of what is possible (see Figure 10–7).

Choosing which iE protocol to use is not as difficult as deciding to move to iE in the first place. As has been shown, Ethernet, TCP/IP, and UDP/IP are capable of delivering data anywhere, anytime. But it is up to the business team to decide if that destination is worth reaching—and then up to the team of application developers, network designers, and control-system integrators to get you there. What follows is a road map of options.

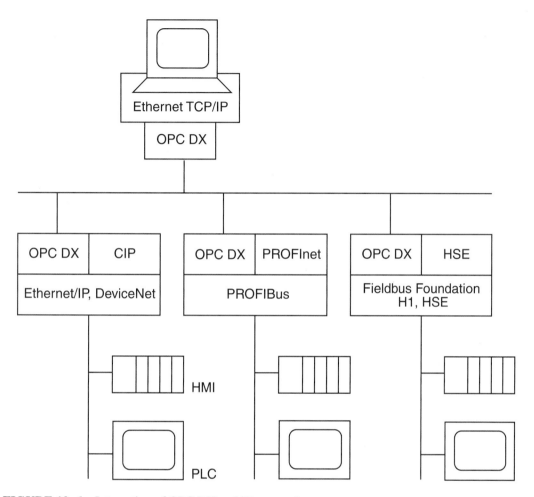

FIGURE 10–6   Integration of OPC DX and iE protocols

FOR EXISTING INSTALLATIONS WITH A SINGLE FIELDBUS
   If the use of a legacy fieldbus will be continued: Deploy a compatible iE version, for example, Modbus and Modbus/TCP, or ControlNet and Ethernet/IP.

   To realize the desired level of data exchange or real-time information flow, middleware software or hardware gateways may be needed. For example, applying a gateway with Modbus/TCP will enable transparent access to the data over an intranet or the Internet, whether for a Web browser or database.

   If changing business models and manufacturing strategies makes a legacy fieldbus obsolete: A fieldbus retrofit may be advisable in conjunction with a data integration initiative.

Control situation	iE application strategy
Existing single fieldbus	
Extend legacy fieldbus, e.g.: Modbus	Add compatible iE modules, gateways, and apps, e.g.: Modbus/TCP, Ethernet/IP
Inadequate legacy fieldbus: Data exchange problems	Apply data exchange solutions using gateways, middleware, and portals, e.g., DataSweep, Lighthammer
Poor network performance	Retrofit full iE protocol with object-model, pub/sub features, e.g.: PROFINet
Internet inaccessibility through firewalls	Replace COM/DCOM with SOAP Add HTTP gateway or portal middleware
Existing multiple fieldbuses	
Combination of DeviceNet, ControlNet, Ethernet IP, PROFIBUS, PROFInet, Foundation Fieldbus H1, HSE	Apply OPC DX
Other combination	Apply gateways, integrating middleware, and portals
Greenfield with no fieldbus	
For simple discrete operations	Apply embedded Modbus/TCP devices
For complex operations/processes	Apply full iE protocol with object-model, pub/sub features, e.g.: PROFInet
For Internet accessibility through firewalls	Apply SOAP-based iE protocol Apply embedded TCP/IP-XML devices, e.g.: OPTO 22

FIGURE 10–7   Control situation and iE application strategy

For example, integration solutions may involve adding gateways, servers, and data-conversion software, such as Datasweep, or Web portals, such as the Lighthammer Illuminator. To access more information, iE protocols may be used that employ a complete object-oriented, pub/sub protocol, such as PROFINet. Or where PC-based control applications are used, it may be possible to move simply from COM and OPC to SOAP and OPC DX. In the case of re-engineering a plant for cellular manufacturing, an optical fiber network may be required to provide the bandwidth to handle the message load.

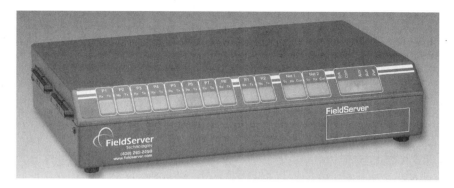

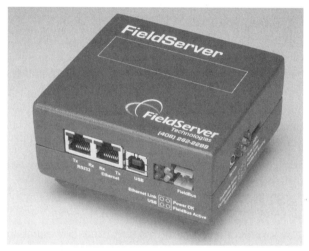

FIGURE 10–8   Ethernet protocol translator (*top*) and gateway.
(Courtesy of FieldServer Technologies.)

IN EXISTING INSTALLATIONS WITH MULTIPLE FIELDBUSES
>   *If any combination of DeviceNet/ControlNet/Ethernet IP, PROFIBUS/PROFInet,
>   Foundation Fieldbus H1/HSE is employed:* Consider OPC DX middleware.
>   *If a number of fieldbuses are employed that are not OPC DX compatible:* Consider
>   protocol translators and gateways to enable data exchange (Figure 10–8).

IN GREENFIELD INSTALLATIONS
>   *If a simple discrete iE solution is possible:* Deploy Modbus/TCP with adequate provisions
>   for real-time performance. If the simplicity of plain-language device and data tags are
>   preferred, then XML-based middleware or .NET for Manufacturing are solutions.
>   *If an event-driven, object- or agent-oriented distributed application architecture is
>   involved:* Select an iE protocol with the appropriate pub/sub capabilities.
>   *If Internet accessibility is the primary consideration:* Choose a SOAP-based iE protocol.

SUMMARY

- An iE protocol's encapsulation technique indicates how closely it is tied to its predecessor fieldbus protocol.
- An iE protocol's messaging model indicates the protocol's flexibility and scalability as the business grows.
- An iE protocol should support profiles and objects if distributed applications and management tools are being developed.
- The simplicity of Modbus/TCP encapsulation makes it the most prevalent iE protocol.
- PROFInet can be implemented on any real-time operating system that can handle a TCP/IP stack and common DCOM and RPC protocols.
- INTERBUS divides up the datagram for transmission over a 100-byte INTERBUS frame field known as a *parameter channel*.
- MMS must be implemented on top of a reliable network and transport mechanism.
- .NET for Manufacturing is Microsoft's answer to DCOM's problems with firewalls.
- CIP allows Ethernet/IP to handle large messages and small I/O data.
- Foundation Fieldbus HSE operates at 100 Mb/s.
- ADS-net employs a bus architecture for pub/sub transactions.
- OPC DX is an integrating interface for interfaces.

? REVIEW QUESTIONS

1. What does an iE protocol's choice of network/transport mechanism tell you?
2. In what layer do profiles and objects reside?
3. What is the advantage of a protocol supporting a full object model?
4. When was the first iE protocol introduced?
5. Why is a layer 3 router advisable when using pub/sub protocols?
6. What transport protocol does PROFInet use?
7. What port must be opened to handle INTERBUS message transfers?
8. What application layer protocol runs underneath SOAP?
9. What protocol is encapsulated inside Ethernet/IP?
10. What transport layer protocols are used by Foundation Fieldbus HSE?
11. Why is the development of the iDA protocol eagerly anticipated?
12. What three iE protocols announced support for OPC DX?

An Open and Closed Case for iE Network Security

11

OBJECTIVES

After reading this chapter, you will be able to:

- Understand the goals of factory-network security
- Appreciate the value of following an established network-security model
- Identify the nine steps used to hack into a system
- Learn how proxy servers and firewalls protect against intrusion
- Understand the role of a network security manager in creating a secure environment

A disgruntled employee, a ruthless competitor, and an industrial spy pose far greater risks to a factory network than the proverbial teenage hacker. They have more knowledge about a plant's operations and control network and, therefore, can do more damage. Securing an iE network presents unique, but not daunting challenges. Highly secure networks are working successfully in many industries. This chapter briefly looks at security goals, plans, problems, and solutions applicable to iE networks that will keep data, information, and operations safe from an attack.

A MENTALITY FOR NETWORK SECURITY

Security at a manufacturing facility usually means a guard at the gate and an ID key at the door. The goal is to prevent the physical removal of property, including documents, computer disks, laptops, as well as prevent tampering with equipment.

Until recently, the possibility of a hacker compromising a factory network was not an issue. That is because either the plant and automation control system was not connected to the outside world, or it was practically inaccessible because of its closed, proprietary design (Figure 11–1).

With the introduction of iE on the factory floor, however, the perspective on industrial network security has changed.

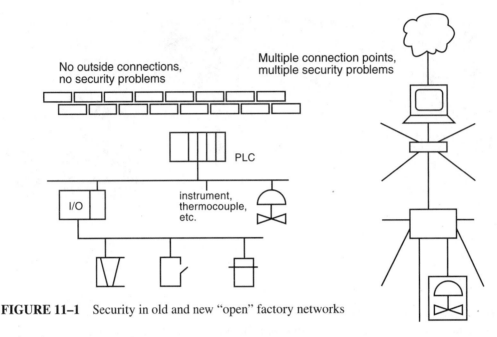

FIGURE 11–1   Security in old and new "open" factory networks

Transparent Factories

Control engineers and network managers must now deal with an "open" networking technology that creates a "transparent factory." This marketing jargon may inspire corporate-level executives dreaming of collaborative business solutions. But to factory and corporate IT managers, it inspires fear. Network managers ask: "Transparent to whom?"

Transparency is a good thing when it allows interconnection between corporate ERP systems, remote monitoring stations, and trading exchanges. It is bad when it provides an entry point to an intruder who can crash a PC-based control network or rip off intellectual assets from a CAD system.

To a plant management's way of thinking, transparency is not worth the risk. A trillion successful communication exchanges over an iE network are insignificant compared to the damage of a production line going down due to a single hacker exploit.

The goal of iE network security, therefore, is zero admittance to unauthorized users. The way to reach that goal is to create a barrier high enough that would-be intruders can not hack into the network anonymously. Anonymity is an essential aspect of the hacker mentality. It is a mind game. And you win if the intruder realizes he must resort to physical entry to get information or cause damage, thereby risking apprehension by plant security officers.

Plan for Success

Planning for network security should be based on a clearly communicated methodology to ensure consistent administration. There are several professional security-planning methodologies available. The SEMATECH model has been developed specifically for the semicon-

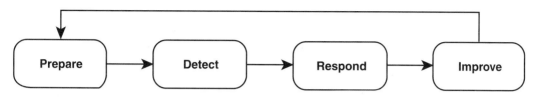

FIGURE 11–2   CERT Network Security Practices Model

ductor industry (www.sematech.org). There is also the "Prepare-Detect-Respond-Improve" model promoted by the CERT Coordination Center (CERT/CC) (Figure 11–2).

CERT NETWORK SECURITY MODEL
CERT/CC (www.cert.org) was established in 1988, and operates out of the Software Engineering Institute at Carnegie Mellon University, in Pittsburgh, Pennsylvania. It grew from a small computer emergency response team backed by the Defense Advanced Research Projects Agency (DARPA) and is still funded by the U.S. Department of Defense and a number of federal civil agencies. It works with the Internet Security Alliance (ISA) and Electronic Industries Alliance (EIA) to coordinate communication among experts of security issues across the Internet community. Since establishing a network security plan is a painful political process, it makes it easier when you can say you are just following CERT/CC standards.

Another planning resource is current information about hacker exploits, viruses, and worms. Such information is available through CERT, anti-virus software companies such as McAfee (www.mcafee.com) or Norton (www.norton.com), as well as books, such as *Hacking Exposed* (Osborne/McGraw-Hill).

Lastly, full documentation on your gateway equipment—switches, hubs, routers, firewalls, and proxy servers—is essential to ensure all equipment is properly configured for utmost security.

CERT/CC acknowledges right up-front that there is no single product, no turn-key solution that prevents security problems. Rather, it takes a *process* to establish safe network practices. Since a description of those practices is available from other sources, the following will focus on defending perimeter gateway routers and the corporate firewall since they are subject to the most serious attacks.

A Portcullis for Packets

In medieval castles, the gate was a logical point of attack. Unlike solid walls with parapets and castellations, it was designed with weaker hinged doorways made from wood or metal. A successful attack—whether a frontal assault or the work of inside betrayal—gave immediate entrance. A successful defense forced the intruders to expose themselves by climbing over or digging under the walls, or camping outside in a siege, where they would be vulnerable to sorties or attacks from allies.

The main tactic in network security is to harden gateway devices to the extent that intruders must put their person at risk by using more traditional means of breaking and entering. That usually means employing a fortified gateway devices, known as **firewalls,** between the business LAN and the industrial-control LAN and possibly between the central network's control and I/O levels.

Firewalls incorporate high-speed microprocessors with the intelligence to distinguish protocol control information in data packets. Properly addressed packets are forwarded through the firewall; improper packets are dropped. Routers and server hardware can function as firewalls. Because routers regulate traffic on the basis of IP address, subnets, protocols, and port numbers, they can completely block packets that don't meet the rules.

Firewalls can be configured as an application proxy server or packet filtering gateway. An application proxy is considered the stronger type of firewall. It inspects the complete data packet. If the payload is valid, it allows the packet to pass. This is a safe, but cumbersome form of protection, because each new application requires writing new rules to handle different packets.

STATIC PACKET FILTERING
Static packet filtering simply looks at the source, destination, and port fields in the TCP (or UDP or ICMP) header.

DYNAMIC PACKET FILTERING
Dynamic packet filtering with **stateful inspection** is a superior form of filtering. Like an application proxy, it uses validation based on whether the packet's contents correspond to the beginning, reply, or request state of the communication session. As shown in Figure 11–3, stateful inspection ensures incoming messages were propagated in response to the client's request, rather than being created as an externally initiated intrusion.

When the firewall is properly designed, configured, and maintained, it is a strong defense that attackers will sidestep. Other forms of attack may be possible—such as worms, viruses, denial of service—and which can be fought off by other means. But if the firewall is solid, the attacker will find anonymous intrusion extremely difficult, if not impossible.

Storming the Gateway

The firewall configuration shown in Figure 11–3 makes it easy for a corporate ERP system to receive an XML page generated from data stored in a manufacturing database. The corporate PC's application makes a request to the server's port 80, and HTTP responds by transmitting over the Internet a data packet containing our valuable number 98.

Unfortunately the system is still open to attack, because improper configuration of certain devices makes intrusion relatively easy.

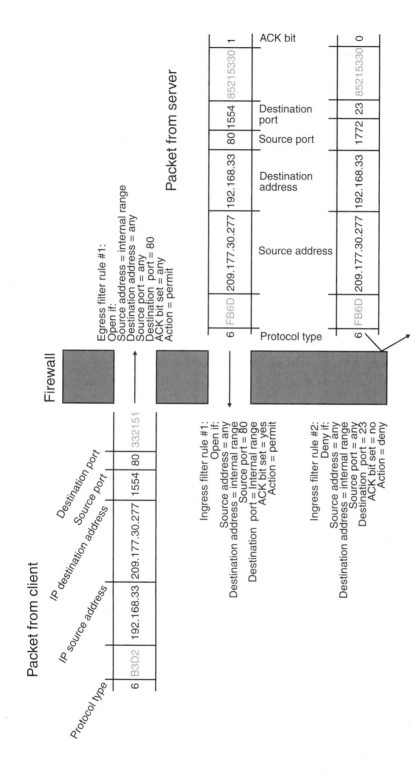

FIGURE 11–3 Packet-filtering firewall

251

BASIC HACKING METHODOLOGY

For example, consider the nine steps involved in hacking into a system:

1. Profiling the Target
   a. Find the target company's Internet server IP number from www.networksolutions.com
   b. Use the PC's "tracert" command to find the IP address of the gateway router
2. Scanning
   a. Use a port probe to discover the router's Telnet port 23 has been left open to allow the system administrator remote access to the router over the Internet
3. Cracking
   a. Establish a Telnet connection with the router, bringing up a password screen
   b. Use a dictionary of manufacturer's default passwords to crack into the router; in this case the default password was simply 1234
4. Penetration
   a. Change the router's settings to allow access to ports used by NetBIOS, indicating a Windows NT Server is behind the firewall
   b. Use NetBIOS to establish a "null session" as explained in numerous books on hacking Windows NT
5. Pilfering
   a. Use Find Computer on the now vulnerable PC to open up the targeted network in Network Neighborhood. Click on the folder bringing up the Enter Network Password dialog box
   b. Guess the password to a local user account—safely assumed to be a common default
   c. Search the root directory for a copy of pcAnywhere, installed by the computer manufacturer or system administrator to perform PC diagnostics from a convenient, remote location. Upload a user profile to use pcAnywhere to gain access to the computer, accessing the entire Network Neighborhood
   d. Access information on network devices as desired
6. Cracking the Administrator's Systems
   a. Use information obtained to crack the system administrator's password
7. Full Penetration
   a. Gain administration privileges to the network
   b. Access administration passwords for servers
   c. Log on to the target database server
8. Reaching the Goal
   a. Access the desired database. Scan database tables. Identify table containing the valuable number 98 and download it
9. Exit and Cover Tracks

The security breech in this example could have been entirely avoided by following a Prepare-Detect-Respond-Improve model to ensure proper device configuration.

Know Thy Enemy

The intruder's main objective is to find a gateway device with a port that is open and ready to service an application. In this case, the attack begins with the discovery that router port 23 is open and ready to host a Telnet session. Hits on a property configured router's Telenet port gives the following result:

Unauthorized Access is Strictly Prohibited!!! The System Administration has been notified. Logoff Immediately!!!

User Access Verification:

Password:
Password:
Password:
%Password: timeout expired!
%Bad passwords

A system administrator can also configure the router to deny external connection to all ports except those providing services to necessary inbound traffic, such as port 80 (providing Web services) and port 443 (providing secure HTTP services).

This simple step would have stopped the intruder cold. Several other simple measures make intrusion difficult.

STRONG PASSWORDS

The firewall router's log-on should use a **strong password,** which is a mix of ASCII upper- and lowercase characters, plus numbers and symbols, at least eight characters long.

All log-ons in Windows systems behind the firewall should use strong passwords and be properly configured against null sessions.

PROPERLY CONFIGURE REMOTE CONNECTION APPLICATIONS

pcAnywhere, Windows Terminal Services, Timbuktu Pro, or VNC should be properly configured. For example, pcAnywhere and other applications open different ports other than 80. Firewalls must be properly configured for messaging applications, such as AOL Instant Messenger, MSN Instant Messenger, ICQ, and Microsoft Exchange, which use a range of ports for some of its services and upon restart might use different ones.

Use strong passwords.   With pcAnywhere, the application can be configured to allow only Specify Individual Caller Privileges, to use strong (case sensitive) passwords, to log off after call completion, to limit log on attempts, to lock out failed log ons, to encrypt session traffic, and to change the default listen port. Finally, strong passwords should be used to protect pcAnywhere profile and setup files.

Resetting port assignments.   In other cases, such as with Exchange, the default port assignment can be changed in the Windows registry. Changing from the expected port creates security via obscurity. The network administrator simply adds the new port to the firewall's **access control list (ACL)** to allow authorized Exchange traffic.

Blocking login.   With pesky messaging applications, changing or blocking the default listening port usually achieves nothing. For example, if your firewall blocks AOL Instant Messenger from receiving messages on port 5190, IM will automatically open another port from 1024 to 65535. A hacker can then run a port scanner to identify and exploit the open port. Consequently, many network administrators will prevent messaging applications from starting a session in the first place by blocking outbound traffic directed to the app's login server.

Such measures force the hacker to attack the system from outside the firewall. Unfortunately, two tactics will still preserve the hacker's anonymity.

Packet Sniffing

Sniffers are legitimate tools used to troubleshoot networking problems. The packet analyzed in Chapter 1 was captured by a sniffer. When installed on a PC, the sniffer application can set the NIC card into **promiscuous mode,** enabling it to accept all the broadcast traffic within the system's collision domain.

A sniffer placed on a backbone or network aggregation point has full access to all traffic. For example, a sniffer installed on a router at the company's ISP can capture the data packet broadcast over the Internet containing the valuable number 98. Thus, the hacker gets the desired information without all the trouble of an intrusive attack.

Sniffers can be counteracted by using switched hubs to route broadcasts only to the specific host, plus network traffic can be encrypted. On internal networks, software can be used to detect NICs in promiscuous mode—essentially sniffing out the sniffer.

IP Spoofing

Spoofing involves redirecting traffic to another device before it reaches its intended destination. Rerouted traffic can then be read by a packet sniffer and forwarded without anyone being the wiser. This attack relies on sniffing out ARP or RIP information, which is then used to send spurious ARP or RIP packets to a switch or router, causing them to reroute traffic.

IP spoofing can be stopped by changing ARP and RIP settings. Another countermeasure is to use virtual private network (VPN) technology to create a tunnel between hosts that encrypts the traffic as it travels from point to point. VPNs can be constructed using Microsoft's point-to-point tunneling protocol (PPTP), layer 2 tunneling protocol (L2TP), and IP security (IPSec) standards. This chapter does not elaborate on VPN technologies nor encryption standards, because their use is usually determined by corporate IT policies outside the plant.

THE ILLOGIC OF NETWORK SECURITY

Ironically, iE security requires closing down an inherently open network technology. In terms of a castle metaphor, instead of building a strong gate, we are constructing a wall that includes secret postern gates and tunnels that a select few can use.

Reconciling Competing Needs

The trouble is that closing ports on a firewall router cuts off a range of services—e-mail, file transfers, remote monitoring—that were partly the reason to deploy iE in the first place. Thus in large enterprises, system administrators must carefully evaluate usage and deploy firewalls accordingly. Unfortunately, evaluation and reconfiguration take a lot of work. Consequently, system administrators are usually very reluctant to activate new services and reconfigure firewalls.

Is there any way to reconcile the competing needs of openness and security? One answer is to use some sort of authentication scheme. Noted high-tech pundit Robert X. Cringely explains:

> We could implement a secure user identity system precisely like telephone Caller ID. It would essentially be an Internet ID. All Internet transactions could be based on it. Anyone who sends me e-mail can be identified. Anything I send can be traced to me. People wouldn't be forced to participate, but if they remain anonymous, I might choose to block them. . . . I think the Internet needs a fingerprint. It does not have to have personal information, but if you break the law it can be traced to you.

He goes on to posit a scenario in which Microsoft issues a modified TCP/IP address to each user via an authentication system, such as Microsoft Passport. This is not just another evil Microsoft scheme. Individualized TCP/IP assignment could also be accomplished through Sun's Liberty or the Free Software Foundation's DotGNU authentication.

It is not a long stretch to imagine personal authentication evolving into something like an individualized TPC/IP number. It could, for example, be a matter of adding an employee-number subset to the current universal description, discovery and integration (UDDI) business registry node number (see www.uddi.org for details). That number would be included in a field to provide the personal identification Cringely describes. In this scenario, one of the privileges of employment would be an individualized TCP/IP number that gives the employee access to business networks. Unfortunately this approach is just one step away from abuse where, for example, denying a person the number cuts them off from personal network access, making employment in an information-based economy impossible.

Perhaps a reasonable alternative is to simply reduce the need for authentication by calibrating more carefully the intellectual assets that need safeguarding. Putting those "secrets" in the strongest possible lockbox will allow all other information to flow freely throughout the organization without making system administrators crack down like fascists or run around like firemen.

Standing Guard

Having built a castle wall and secured the portals, it is tempting to assume that no one needs to stand guard. After all, you've deployed state-of-the-art firewalls and switches—why involve a network security manager?

The reason: People are the cause of security problems. It takes people to create security solutions. As with VPN deployment mentioned earlier, security management for a factory network may be the responsibility of corporate IT. But depending on the industry, that responsibility may fall under a plant IT administrator.

According to CERT, network security practices involve five general steps:

1. Harden and secure your systems by establishing secure configurations
2. Prepare for intrusions by getting ready for detection and response
3. Detect intrusion quickly
4. Respond to intrusions to minimize damage
5. Improve security to help protect against future attacks

Taking the first step is the most important. That is because properly configured and updated hardware and software create a high barrier by eliminating the most obvious exploits. A sophisticated hacker or an industrial spy may persevere, however. The prepared network administrator will employ early warning software, such as port scanning detectors and network usage logs, to detect activities that may signal an attempted intrusion.

The battle of wits between hackers and network administrators is the subject of books and movies. But in real life, security problems are not usually the result of exotic exploits but common negligence. Here are some common ways networks are brought down:

1. Downloading a file without virus checking turned on.
2. Using obvious network log in passwords or no password at all.
3. Using "plug and play" devices that choke factory-networks bandwidth by creating a broadcast storm when advertising its address on the network.
4. Using a faulty NIC that spits malformed packets (runts) onto the network (a switch will contain this problem, while a hub lets these packets pass).
5. Replacing or adding devices that use a duplicate IP address.
6. Assuming the default settings of newly purchased iE products are secure.
7. Running older software with unpatched network vulnerabilities.

And there are too many cases where accessing industrial secrets is simply a matter of stealing a laptop or desktop PC containing unencrypted information on the hard drive. In network security, a little bit of intelligence goes a long way—if we only use it.

SUMMARY

- In the past, hacking into industrial networks was not an issue, because it was not connected to the outside world or was inaccessible due to its proprietary design.
- The goal of iE network security is zero admittance to unauthorized users.
- Full documentation on your gateway equipment is essential to ensure equipment is properly configured for security.
- The objective of the intruder is to find a gateway device with a port that is open and ready to service an application.
- A strong password is a mix of ASCII upper- and lowercase characters, plus numbers and symbols, at least eight characters long.
- With pesky messaging applications, changing or blocking the default listening port usually achieves nothing.
- A sniffer placed on a backbone or network aggregation point has full access to all traffic.
- Spoofing involves redirecting traffic to another device before it reaches its intended destination.
- A reasonable alternative to stifling user authentication schemes is to calibrate more carefully the intellectual assets that need safeguarding.
- People are the cause of security problems. It takes people to create security solutions.

? REVIEW QUESTIONS

1. Why are *open* factory networks more prone to security problems?
2. What are the four parts of the CERT Network Security Practices Model, and which is most important?
3. What is the difference between static and dynamic packet filtering?
4. Explain the role of a firewall in stopping the hacking process.
5. Why is proper configuration of peripherals (routers) and core (PCs) equipment important?
6. Why do remote access applications create security vulnerabilities?
7. What is one disadvantage of a password and user authentication scheme?
8. Why is technology a necessary, but not an entirely effective safeguard, in creating a secure network environment?
9. What are two common everyday security problems created by users?

APPENDICES

Appendix A

Putting iE Technologies to Work: A Brief Planning Guide for Network Design Teams

The subject of IP network planning can be complex, as shown by IBM's 324-page *IP Network Design Guide* (see www.redbooks.ibm.com/redbooks/SG242580.html). Planning an iE network for process control and factory automation is even more complex. The purpose of this guide, therefore, is to provide a simple overview of the roles played by control engineers, systems integrators, network administrators, and information architects in planning such a network. Success depends on their ability to function as a team (see Figure A–1). This is especially important between the control engineer and the network administrator, whose respective interests in operational reliability and enterprise security should be complementary rather than a source of conflict.

1. Forming the Team

A. RESPONSIBILITIES

Control engineer (CE). Responsible for designing, implementing, and maintaining the control system. Responsible for plant-side network security.

Systems integrator (SI). Demonstrates that the system complies with the specifications. Checks performance requirements—response time, data points—and appraises cost. Configures devices with correct IP-addresses.

Network administrator (NA). Provides a physical connection (e.g., 100Base-T connection) and logical connection (e.g., an IP address, a subnet mask, and a default gateway/router address) to the enterprise system. Responsible for enterprise-side network security.

Information architect (IA). Ensures integration of back-end applications with the data being made available from factory processes and operations.

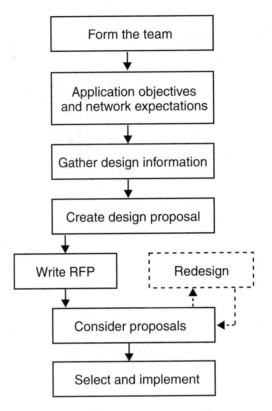

FIGURE A–1   Network-design planning flow

B. VENDOR SELECTION (CE/NA)

As much as possible, the CE and NA should work together to select equipment vendors, solution providers, and system integrators. Making choices that apply to both the plant and enterprise promotes compatibility and common support practices. Hybrid, "best-of- breed" equipment selections should be tested together to ensure smooth deployment.

2. Application Objectives and Network Expectations (CE/NA)

The team needs to adopt a common planning approach in order to reach consensus about objectives and expectations. A top-down planning approach is very useful. As shown in this book, operational and business requirements dictate the kinds of applications that will be employed over the network (Chapters 2 and 3). Applications, in turn, dictate upper-layer protocols (Chapters 9 and 10), which then require certain network resources (Chapter 6) and physical components (Chapter 5). More specifically, a top-down planning method starts by considering user application and application-layer issues, then moves down to transport, network, data link, and physical requirements. Note that this approach resembles the seven-layer communication stack—an easily understood reference point for team planning.

Network expectations will be based on a number of factors, which include:

- Number and type of users (manufacturing cells, facility campus, remote locations, business enterprise, extended enterprise, value chain)
- Number and kinds of factory productivity applications (HMI, MES, MRP, SCADA)
- Operational network demands (block I/O transfers, programming/recipes transfers, and time-sensitive interlocking, polling, cyclic "heartbeat," and report-on-event messaging)
- Integration with business applications (SAP, ERP, APS, DBMS), with collaboration solutions (CAD/CAM/CAE, PDM), and with value-chain applications (VMI, SCM, CRM)
- Application-layer services (HTTP, HTTPS, encryption, real-time, publish/subscribe, RPC, through firewall)
- Security (SSL for TCP versus encryption for UDP, proper procedures and configuration of firewall rules and procedures)
- Performance (throughput and response time)
- Economics (cost of migration, components, cabling, installation, validation, support)
- Network and bandwidth management (passive techniques, such as creating collision domains and subnets to isolate bulk traffic from I/O data, limiting plug-and-play devices, shortening cable lengths; active techniques, such as store-and-forward hubs, network analyzers, bandwidth managers)
- Availability and reliability (fault tolerance, redundancy, QoS)
- Scalability and modularity (flat or hierarchical topology, spare ports on switches, IP address allocation, directory needs)
- Geography (length of cables, wireless IP, remote access)

These factors also help determine the best tactics for iE network deployment. As discussed in Chapter 10, Ethernet can be applied to the plant network in four ways:

1. Completely supersede the existing fieldbus
2. Extend the existing control network interface by using an iE version of the fieldbus
3. Use Ethernet as a backbone with drops connecting, for example, TCP/IP communication modules in PLCs and network interfaces in I/O modules
4. Use a gateway between the legacy fieldbus network and the Ethernet network

3. Gather Design Information

The actual network design should be based on careful research by team members, which can be delegated as follows:

A. ANALYZE PERFORMANCE (CE)

Calculate latency and throughput for intended devices. Latency is calculated for each component; throughput is calculated for the system.

Err toward conserving bandwidth. Partition the network with routers and switches as necessary. If more than two routers are required, use a backbone architecture. A backbone allows network segmentation and cross-segment traffic, which minimizes latency. Data throughput requirements may necessitate Fast Ethernet (100 Mbps).

B. SITE ENVIRONMENT (CE)

Determine environmental requirements. Temperature, chemicals, spark, liquid, dust, vibration, electromagnetic interference. Consider IP67 sealed RJ45 jacks, screened or shielded cables.

Determine physical constraints. Distances (100-meter limitations), cable routing. Use bulkheads through enclosures.

C. QUANTIFY COMPONENTS (CE/SI)

Segments, ports (one for each segment), switches, patch panels (to interface I/O devices, switches, or hubs), and bulkhead enclosures.

D. CABLING (CE/SI)

Types. Shielded (STP), screened (ScTP) cable, unshielded (UTP) cable, oil-resistant jackets, plenum-rated cables, weld-splatter proofing cables, fiber optics.

Consider Category 5i UTP with moderate and ScP with high electrical-noise immunity. Install four-pair cable where possible—power can be added on unused pairs in pending 802.3af standards. Consider single point grounds (grounding equipment will ground RJ45 shielded cable).

Avoid runs near lights, motors, drive controllers, arc welders, and conduits. Cross power lines at right angles. Run at least 1.5 m (5 ft) from high voltage. In conduit, only use shielded cable. Stay at least 3 inches from electric light and power conductors; 5, 12, and 24 inches from power sources under 2, from 2 through 5, and over 5 kVA, respectively. Bend cable no less than four times its diameter.

E. CONNECTORS (CE/SI)

Consider IP67 sealed RJ45 connectors. Apply appropriate terminations (point-to-point connections must end in a null termination). Use modular cable connections or punch-down methods. Match connector/cable impedance. Select fiber SC, ST, MT-RJ, or LC connectors.

4. Create Design Proposal (Team)

Designing a LAN for a small factory with a few nodes is far different from building one with thousands of enterprise users. In building a small plant network, a flat design can be used to connect all devices within one subnet for exchanging recipes and programming (Figure C–1), or switches can be used to enable real-time data acquisition (Figure C–2), or a router can be used to access operational data over the Internet (Figure C–4). For a large operation with many users and many kinds of data transactions, such as an automotive manufacturing plant, a hierarchical network design is used (Figure C–5). (For network-design software tools to assist in designing new and modifying existing networks, see www.nwfusion.com/reviews/2001/1203rev.html).

For small installations, such as remote monitoring or machine control, connect a compatible iE interface in a PLC to communicate with devices over a legacy network. It may also be feasible to simply replace devices with embedded TCP/IP interface and reconfigure the system from PLC to PC-based control.

For larger distributed control systems (CE) requiring a continuous network connection, build an Ethernet backbone with switches isolating segments requiring real-time data transfer among iE-enabled devices, I/Os, and controllers.

For enterprises (IA) that do not require a constant connection, firewall restrictions can be handled by proxy servers and gateways. Data residing in different silos may require middleware development and data integration. Web portals can be used to facilitate accessing information over the Internet.

B. WRITING THE RFP (CE/NA)

Ideally, the control engineer and network administrator should work together to develop consistent RFP priorities for both the plant and enterprise. Attention should be paid to writing specific contract language for network purchases. A poorly written specification is likely to result in poor network performance, necessitating costly rescue work and startup delays.

According to Victor Wegelin of PMA Concepts, "I find a general weakness is inadequately specifying network performance in the original scope of work, so that in the end, the network fails to meet expectations. Therefore, I prefer to see guidelines that require attenuation profiles (hardware test) and bandwidth profile (loading) of the network. This requires the contractor to supply documentation showing industry-accepted test methods, such as TDR (time-domain-reflectometry test for cable integrity) and software benchmarking as conditions of project acceptance. Many such tools are in wide acceptance today. Vendors who object to such stringent requirements should not be invited to bid."

The bottom line: All prime contractors should be interviewed on overall strategies, methodologies, experience, and project-acceptance criteria.

6. Implementation Roles

Other members of the team—such as a trusted control system integrator or solution provider—may be involved in analyzing application needs and collecting design information. But after a design has been selected and implementation begins, these roles become important:

A. PROJECT MANAGEMENT (SI)

The integrator initiates project management, reviews estimate versus scope, identifies project management team, creates project schedule, identifies resources, details work, and provides submittals, warranty, and support.

B. TESTING (SI)
Devise test documentation requirements/test plan, factory test, and site test.

C. INSTALLATION AND IMPLEMENTATION (SI/CE/NA)
Construction must meet worksite safety, neatness, and security requirements. Define responsibilities of integrator, client, and subcontractors.

D. SITE STARTUP (SI)
Develop startup plan, perform startup, and generate documents.

E. SUPPORT (SI) AND TRAINING (SI/CE/NA)
Develop training procedures and troubleshooting response plans.

Appendix B
Practical IP Addressing and Subnet Masking

SITUATION: GETTING A BUSINESS LAPTOP TO COMMUNICATE ON A FACTORY NETWORK

When a laptop computer from an office is plugged into a factory network, chances are it won't communicate. Why?

- It could be the cable. Check to see if one light-emitting diode (LED) on the interface is illuminated. This indicates a physical link has been established (a TX LED indicates transmitting data, a RX LED indicates receiving data).
- If the cable is okay, the problem is higher up the protocol stack—the laptop probably doesn't have the proper IP address.

Devices on a TCP/IP network communicate by sending packets to specific IP addresses. The laptop must have a compatible IP address or it will not be recognized by other hosts on the network and, therefore, be unable to transfer messages.

Devices with compatible IP addresses can send messages between each other without using an intervening gateway. They are said to be on the same **subnet.** A **subnet mask** is a specific number used with the IPv4 addressing scheme to create IP addresses that are compatible for a given subnet. The subnet mask delineates the portion of the IP address that will be used for the network address from that used for the host address.

To better understand the function of subnet masking, consider the laptop example. Before the laptop sends an IP packet to another device, the laptop may check its own internal table of addresses (local host table) or a domain name server (DNS) to determine if that host is on the same subnet. If there is a match, TCP/IP simply addresses the packet to that host and transmits the packet. If there is no match, the destination host is assumed to be outside the subnet. The originating device looks up the default gateway address and attaches it to the packet, which is then transmitted. The destination gateway (such as a router) then forwards the packet to the next network.

In our example, when the office laptop is plugged into the factory network, it can't communicate because it can't match the IP address of the hosts on the local network (their addresses aren't in its local host table). Nor does it know the address of a gateway router through which it could reach a DNS to obtain addresses.

To solve this problem, the laptop needs to be reconfigured with an IP address that's compatible with the factory network.

HOW IP ADDRESSING WORKS

In the very early days of the Internet, proper configuration was a matter of selecting a unique, 32-bit Internet address from the proper class of addresses. With the technique known as Classful IP addressing, the IP address was divided into two parts: the network address (now known as the network prefix) and the host number.

All hosts on the same network share the same network prefix but have a unique host address.

The trouble is that the 32-bit address is one continuous string of binary numbers that make it difficult to differentiate the portion of the IP address representing the network and the part representing the host.

To better understand the problem, consider this street address:

First and 10 Corporation
123 Fourth Street
Century, AK 00234

Purely numerically, it looks like this:

110, 123, 4, 100, 00234

But because the zip code represents the city and state, the information required for delivery can be simplified to:

110, 123, 4, 00234

The last group of four numbers represents the delivery area; the front three groups defines the individual recipient.

The IP address system works the same way, except the order of the numbers is reversed:

234.4.123.110

The front group of numbers represents the delivery area—or network—and the following groups represent the individual host's address. This number is in decimal form to make it easy for human readers. But translated into the binary numbers that computers read, it looks like this:

234.	4.	123.	110
11101010	00000100	01111011	01101110

For ease of reading, the string of 32 binary digits has been spaced in eight-digit binary words, each known as an **octet** or a **byte,** which correspond with each of the four decimal numbers, known as **quads.**

In machine language, this number is a continuous string of binary digits. How then can a device tell which is the network portion of the address (the zip code and street) and which is the host device's address (the house number)?

In 1981, a technique was developed known as **classful addressing.** This method identifies the dividing point between the network prefix and the host number by beginning each address with a self-encoding key. Class A addresses began with the bit "0," reserving the first eight binary digits for the network prefix. Class B began with the bits "1,0," reserving the first 16 binary digits, and Class C began with the bits "1,1,0," reserving the first 24 binary digits.

Classful addressing based on the IPv4 addressing system combines a total of 32 binary digits to provide over 4 billion IP addresses. This proved insufficient. Today the IPv6 scheme is being developed to offer many more addresses by combining 64 binary digits.

THE ROLE OF SUBNET MASKING

In 1985 another method was developed to provide a more flexible addressing scheme than wastefully assigning scarce IP addresses. This technique is known as subnet masking.

With subnet masking, the IP address is no longer in two parts, and there is no self-encoding class key. Rather, it is divided into three parts: network prefix, subnet number, and host number. The first two parts work together as an extended-network prefix. Here is how it functions.

Returning to our IP address example, a subnet number of 255.255.255.0 is used to get the router to read the first three quads (in **bold**) as the network prefix:

234.	**4.**	**123.**	110
11101010	**00000100**	**01111011**	01101110

But how is the subnet number used to mark the first 24 binary digits as the network prefix? The subnet number is itself translated it into a binary number:

255.	255.	255.	0
11111111	11111111	11111111	00000000

Then the router reads every binary digit in the IP address that corresponds with a binary 1 on the subnet address, ignoring those that correspond with a 0. In other words, the subnet number acts as a screen allowing only a portion of the IP address to be interpreted as the network prefix—hence the term subnet "masking."

Here's how the two binary numbers correspond:

			110
11101010	**00000100**	**01111011**	01101110
11111111	**11111111**	**11111111**	00000000
234.	**4.**	**123.**	

Note that the subnet mask only allows the first three quads to be read, not the last quad. Thus, the three quads—234.4.123—signify the network prefix. The number 110 signifies the host address, one of 254 possible hosts on this particular network. (There are usually two fewer hosts than the total possible number.) In this case, network engineers would know that the three quads represent a Class C IP address, which uses only the first three quads for the network prefix.

In the 1990s, class-based addressing became outmoded. A new way to indicate a Class C address was developed that uses the first 24 binaries for the network prefix, which is written as 234.4.123.110/24. This notation is used in routers that support classless inter-domain routing (CIDR), a technique that allows for more flexible and efficient allocation of IPv4 addresses. (Note: Older RIP-1 routers support only classful addressing.)

Thanks to CIDR, a LAN can be more effectively subdivided. For example a subnet mask of 255.255.255.224 tells CIDR routers to read the 3 binary numbers in the fourth quad as part of the LAN address (in **bold**):

255.	255.	255.	224
11111111	11111111	11111111	11100000
11101010	**00000100**	**01111011**	**011**01110

A simple way to designate that the first 27 binaries represent the network prefix is to write 234.4.123.110/27. In this case, there are only five binary digits remaining that can be combined into 32 possible device addresses. Therefore, 30 hosts (remember, two less than the total) can be connected in this subnetwork.

The traditional Class B addresses with a 255.255.0.0 subnet mask can now be represented by /16 to tell the router that the initial two quads—16 binary digits—is the network prefix.

Class A addresses with a 255.0.0.0 subnet mask can now be signified by /8. (Class D addresses are used for multicasting, as discussed in Chapter 9.)

With CIDR, more flexible address combinations are possible than with classful addressing. Plus, CIDR routers can use two numbers, for example, 234.4.123.110/255.255.255.224 to advertise the path to reach one of 32 hosts in the /27 subnet, rather than advertise each IP address individually.

Figure B–1 shows the wider range of IP addresses that can be created with CIDR. To help network administrators manage the task of assigning network and host addresses, such tables show the possible combinations.

Just as duplicate IP addresses must be prevented on the LAN, they must also be avoided on the Internet. You can use any IP address as long as the device is not connected to the Internet. (For example, in a subnet the router will use NAT to swap the host's potentially conflicting IP address with an address legitimate for the Internet.)

If the device is connected to the Internet, it must use an IP address from three ranges reserved for private networks. These addresses will be forwarded by Internet routers:

10.0.0.0 to 10.255.255.255 (10/8)
172.16.0.0 to 172.31.255.255 (172.16/12)
192.168.0.0 to 192.168.255.255 (192.168/16)

IP addresses outside these ranges are usually purchased from an Internet service provider (ISP) or from the American Registry of Internet Numbers (www.arin.net).

CONFIGURING THE IP ADDRESS AND SUBNET MASK

Returning to the preceding laptop problem, a solution is now possible by applying our knowledge of network prefixes, subnet masks, and host addresses.

Assuming the laptop is using a Windows 98 operating system, we can detect its IP address by going to the "Start Menu," selecting "Run," enter "command," and after ">" enter "winipcfg." With Windows 2000, enter "command" and type "ipconfig" after the command line prompt. Do not enter quotation marks. A dialog box shows IP address 10.0.0.125, subnet mask 255.255.0.0, and default gateway 10.0.0.1.

By running the same check on a PC workstation on the factory network, we see the workstation has an IP address of 234.4.123.113, a subnet mask of 255.255.255.224, and a gateway of 234.4.123.110. Armed with this information and the knowledge gained from preceding chapters, we can determine five ways to get the laptop communicating:

1. Reconfigure the laptop's IP addresses and subnet mask manually The most direct solution is to use the tools provided by Windows to reconfigure the laptop's IP address. On Windows 98, assign a private IP address for a simple network and open the Network dialog box (go to Start>Control Panel>Network).

Class /CIDR	Classful IP address range	Subnet mask (CIDR method can use Class B mask with Class C IP addresses)	Number of possible network prefixes	Number of useable host addresses (two fewer than possible)
A /8	1.xxx.xxx.xxx thru 126.xxx.xxx.xxx	255.0.0.0	256	16,777,214
B /16	128.xxx.xxx thru 191.255.xxx.xxx	255.255.0.0	65,536 (256 Class A 1/256 of Class C)	65,534
/17		255.255.128.0	131,072 (2 Class B)	32,766
/18		255.255.192.0	262,144 (4 Class B)	16,382
/19		255.255.224.0	524,288 (8 Class B)	8,190
/20		255.255.240.0	1,048,576 (16 Class B)	4,094
/21		255.255.248.0	2,097,152 (32 Class B)	2,046
/22		255.255.252.0	4,194,304 (64 Class B)	1,022
/23		255.255.254.0	8,388,608 (128 Class B)	510
C /24	192.0.0.xxx thru 223.255.255.xxx	255.255.255.0	16,777,216 (256 Class B)	254
/25		255.255.255.128	33,554,432 (2 Class C)	126
/26		255.255.255.192	67,108,864 (4 Class C)	62
/27		255.255.255.224	134,217,728 (8 Class C)	30

FIGURE B–1   Some subnet mask values used with IP address ranges

Select the TCP/IP Ethernet adapter from the list titled "the following network components are installed." Click Properties. To assign a static IP address, click Specify an IP address, and then type in the IP address 234.4.123.114 (assuming no other device on the subnet uses this number), a subnet mask 255.255.255.224, and a gateway of 234.4.123.110. Using a static IP address prevents the DHCP server from supplying an address dynamically.

2. Employ two separate network interface cards (NICs) in the laptop   Plug the laptop into the factory network using a second NIC that is configured with the IP, subnet, and gateway numbers given above.

3. Configure a single NIC with multiple IP addresses   Configuring multiple IP addresses and subnet masks on one network interface can be done for Windows 2000 (consult Windows "Help") but involves editing the registry in Windows 98–not recommended nor described here.

4. Plug the laptop into a factory LAN subnet with a router   Until now, we have assumed the laptop is using a fixed IP address assigned by a network administrator. But it is also possible (and very likely in an office setting) that the laptop obtained its IP address dynamically. In an office environment, network administrators save trouble by using routers employing dynamic host control protocol (DHCP) to assign IP addresses automatically.

A factory network can be similarly designed with a subnetwork employing a router with DHCP. The laptop (follow procedure in step 1, but select option to obtain an IP address automatically) can then be plugged into a hub within the subnet or directly into a host port in the router. Then on boot up, the laptop will obtain a proper IP address (192.168.0.33), subnet mask (255.255.255.0), gateway address (192.168.0.1), and DNS address from the router. If there are other hosts on the subnet, the laptop will be able to communicate with them directly, because they share the 192.168.0 network prefix established by the subnet mask.

To communicate beyond this subnet—such as a network with a compatible TCP/IP I/O module, PLC, or DSL modem—the router will serve as the gateway. Using a technique known as natural address translation (NAT), the router will replace the IP address in the laptop's packets with its own and forward them to the port interfacing with the 234.4.123.110/27 subnet.

5. Plug a wireless access point (WAP) into the factory network and a wireless card into the laptop   With this solution, the wired interface of the WAP is configured for the factory network. The WAP's wireless interface is configured for the laptop. Then similar to a router, the WAP forwards the laptop's wireless packets onto the factory's wired network by performing DHCP and NAT functions.

A WAP can also work in bridge mode, transmitting packets based on the MAC address rather than the IP address. Bridge mode provides a simple connection between the wired network and the wireless client. This approach presumes, however, that the laptop is configured with a valid subnet and MAC addresses for the hosts on the other side of the bridge. Remember: Hubs, bridges, and switches must connect hosts within the same subnet; routers can connect different subnets.

ADDRESS ALLOCATION

Using proper IP addresses and subnet masks is necessary not only for the individual device but also for the entire network. When allocating addresses for the network, the network engineer will create a table showing blocks of addresses, as shown in Figure B–2. Note that the CIDR addressing method is used. By employing a subnet mask allowing for less than the 24-bit network prefix typical with 192.168.x.x Class C addresses, more host address space is available for future expansion.

Subnet	Number of devices	CIDR address	Start	End	Mask	Number of hosts (subtract 2)
Supervisory	200	192.168.0.0/24	192.168.0.0	192.168.0.255	255.255.255.0	256
Process 1	300	192.168.1.0/23	192.168.1.0	192.168.2.255	255.255.254.0	512
Process 2	300	192.168.3.0/23	192.168.3.0	192.168.4.255	255.255.254.0	512
Line 1	1000	192.168.5.0/22	192.168.5.0	192.168.8.255	255.255.252.0	1024
Line 2	1000	192.168.9.0/22	192.168.9.0	192.168.12.255	255.255.252.0	1024
Line 3	800	192.168.13.0/22	192.168.13.0	192.168.16.255	255.255.252.0	1024
40 Cells	1200	192.168.49.0/21	192.168.49.0	192.168.56.255	255.255.248.0	2048
Cell 1	30	192.168.49.0/27	192.168.49.0	192.168.49.3	255.255.255.224	32
Cell 2, etc.	30	192.168.49.32/27	192.168.49.32	192.168.49.63	255.255.255.224	32

FIGURE B–2   Sample CIDR address allocation for a factory network

UTILITIES TO DETECT ADDRESS AND NETWORK PROBLEMS

TCP/IP supports a number of basic utilities that can be used to detect and solve IP address and network problems:

1. PING   This utility sends a 32-byte test packet to an IP address or URL and, like a sonar signal, measures the packet's bounce-back speed. PING is used to confirm that a device is on the network and able to receive IP packets.

A. From the Windows DOS command line (described above), enter:

        ping xxx.xxx.xxx.xxx (IP address)

or

        ping xxx.xxxxxx.xxx (Website URL)

Note: PING will poll the DNS to translate the URL into the server's IP address.

B. When PING detects the device, it returns information like this:

        Pinging 192.168.1.1 with 32 bytes of data:

        Reply from 192.168.1.1: bytes=32 time<10ms TTL=254
        Reply from 192.168.1.1: bytes=32 time<10ms TTL=254
        Reply from 192.168.1.1: bytes=32 time<10ms TTL=254
        Reply from 192.168.1.1: bytes=32 time<10ms TTL=254

        Ping statistics for 192.168.1.1:
        Packets: Sent = 4, Received = 4, Lost = 0 (0% loss),
        Approximate round trip times in milliseconds:
        Minimum = 0ms, Maximum = 0ms, Average = 0ms

In this case:

time = shows that PING sent four 32-byte test packets, each of which reached the destination and were acknowledged in under 10 milliseconds (ms). Because device 192.168.1.1 is the IP address for the network router's LAN interface, the roundtrip time is fast—under 1 ms, which rounds down to 0 ms. Pinging the IP address for the router's WAN interface (10.0.0.140) would be just as fast. Long roundtrip times would indicate a problem with the router.
TTL = Time to Live indicates the number of times the packet can be retransmitted before being discarded. Each router hop subtracts one number; at zero, the packet is discarded.

C. An index of specific functions (switches) for PING can be accessed from the DOS command just by entering "ping."

2. TRACEROUTE   This utility sends three test packets and displays information on each hop between the sender and the destination. It is used to detect IP addresses and speeds between hops, thereby revealing failure in a complex network by telling you exactly the time it takes for the test packet to travel between hops.

   A. From the Windows DOS command line (described above), enter:

        tracert xxx.xxx.xxx.xxx (IP address)

  or

        tracert xxx.xxxxxx.xxx (Website URL)

   Note: TRACEROUTE uses DNS to translate the URL into the server's IP address.

   B. TRACEROUTE returns information like this:

        tracert 67.107.237.227

        Tracing route to adsl-67-107-237-227.testdomain.com
        [67.107.237.227] over a maximum of 30 hops:

        1 <10 ms <10 ms <10 ms 192.168.1.1
        2 42 ms 55 ms 41 ms adsl-207-44-112-1.testdomain.com
        [207.47.117.7]
        3 adsl-207-47-117-7.testdomain.com [207.47.117.1] reports:
        Destination host unreachable.

        Trace complete.

In this case:

The hop to the router at 192.168.1.1 occurs very fast, as before. But the three roundtrip test packet transmissions to the phone company's digital subscriber line (DSL) switch took varying amounts of time. Note that the switch was configured to prevent PING or TRACERT of the specific IP address.

3. NETSTAT   The NETSTAT utility reports the network status of the local host by polling the ports of active connections.

   A. From the Windows DOS command line (described above), enter:

        netstat

B. NETSTAT returns information like this:

Active Connections

Proto Local Address	Foreign Address	State
TCP station1:1039	t1-66-109-247-097.test.com:1250	ESTABLISHED
TCP station1:1048	t1-66-109-247-097.test.com:1250	ESTABLISHED
TCP station1:1362	t1-66-109-247-097.test.com:1044	ESTABLISHED
TCP station1:1371	10.0.0.1:telnet	ESTABLISHED
TCP station1:1256	www.google.com:80	CLOSE_WAIT

In this case:

NETSTAT reports what protocol a device is using and the status of the active ports, the foreign (destination) device's IP address and port numbers of remote sockets, and the status of the servers. In this case, the first three sockets were created by an e-mail application, in addition to the ports opened by Telnet and HTTP applications.

4. ARP   The ARP utility maps IP address to MAC addresses. This tool can detect if more than one station on the network has the same TCP/IP address by returning two ARP responses for a single ARP command.

A. From the Windows DOS command line (described above), enter:

arp

This will return a list of switches:

ARP -a [inet_addr] [-N if_addr]	
-a	Displays current ARP entries by interrogating the current protocol data. If inet_addr is specified, the IP and Physical addresses for only the specified computer are displayed. If more than one network interface uses ARP, entries for each ARP table are displayed.
-g	Same as -a
inet_addr	Specifies an internet address
-N if_addr	Displays the ARP entries for the network interface specified by if_addr
-d	Deletes the host specified by inet_addr

-s	Adds the host and associates the Internet address inet_addr with the Physical address eth_addr. The Physical address is given as 6 hexadecimal bytes separated by hyphens. The entry is permanent
eth_addr	Specifies a physical address
if_addr	If present, this specifies the Internet address of the interface whose address translation table should be modified. If not present, the first applicable interface will be used.

B. ARP-A returns this information:

    Interface: 192.168.1.33 on Interface 0x1000002
    Internet Address   Physical Address   Type
    192.168.1.1   00-03-fd-67-2d-41   dynamic

In this case, the device's IP address is shown along with the subnet address, MAC address, and a notice that the address was assigned automatically. Because workstation TCP/IP addresses can change, the internal ARP table may timeout and return no information.

Appendix C

Examples of Unsegmented, Switched, and Tiered Networks

EXAMPLE 1: UNSEGMENTED NETWORK, SINGLE SUBNET

This example[1] shows a very basic four-node network used to download instructions to a PLC controlling a punch press (Figure C–1). All plant and offices devices are connected by a hub and are therefore on a single subnet and a single collision domain. Because deterministic, real-time data transmission isn't an issue, the collisions that may occasionally occur are not critical.

EXAMPLE 2: SWITCHED NETWORK, TWO SUBNETS

This example shows a more complex 16-node network segmented into eight collision domains by a switch (Figure C–2). The switch uses a look up to table, which it creates by learning the MAC address of each node in the segment (Figure C–3). The separate collision domains isolate critical PLC traffic from less critical network messaging. Yet data can still flow up to the data historian on a scheduled basis. Note that all factory devices are on the same subnet, but a router is used to create a separate subnet for the office LAN.

---

[1]This example courtesy of Loman Control Systems, Inc., Lititz, Pa.

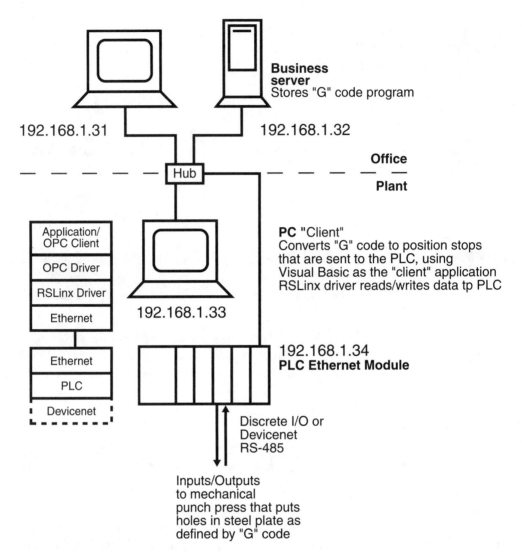

Business server
Stores "G" code program

192.168.1.31

192.168.1.32

Office

Hub

Plant

Application/OPC Client

OPC Driver

RSLinx Driver

Ethernet

192.168.1.33

Ethernet

PLC

Devicenet

PC "Client"
Converts "G" code to position stops that are sent to the PLC, using Visual Basic as the "client" application RSLinx driver reads/writes data tp PLC

192.168.1.34
PLC Ethernet Module

Discrete I/O or Devicenet RS-485

Inputs/Outputs to mechanical punch press that puts holes in steel plate as defined by "G" code

FIGURE C–1   Simple topology for transmitting punch press programming: single machine, single PC, single translation

(Courtesy Loman Control Systems)

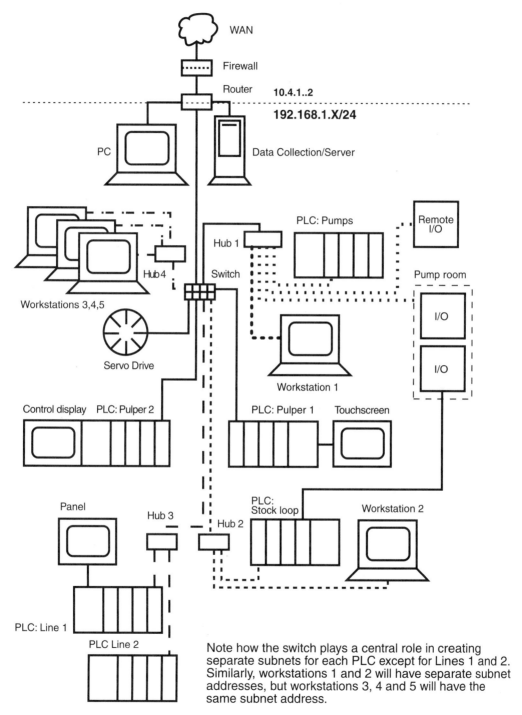

Note how the switch plays a central role in creating separate subnets for each PLC except for Lines 1 and 2. Similarly, workstations 1 and 2 will have separate subnet addresses, but workstations 3, 4 and 5 will have the same subnet address.

FIGURE C–2   Switched topology for controlling a moderately complex application

Node	Device	IP Address	MAC	Switch port	Collision domain
1	Station 1	192.168.1.2	00-AD-00-33-44-76	1	1
2	PLC: Pumps	192.168.1.3	00-44-00-22-88-77	1	
3	Remote I/O	192.168.1.4	00-12-09-33-B4-77	1	
4	Pump Room	192.168.1.5	00-D6-06-36-46-66	1	
5	PLC Pulper 1	192.168.1.6	00-A1-00-12-44-13	2	2
6	PLC Stock	192.168.1.7	00-1D-00-44-44-64	3	3
7	Station 2	192.168.1.8	0B-D1-10-13-14-54	3	
8	PLC Line 1	192.168.1.9	00-GD-G0-36-99-07	4	4
9	PLC Line 2	192.168.1.10	00-DF-0F-33-94-53	4	
10	PLC Pulper 2	192.168.1.11	00-CE-A0-33-88-78	5	5
11	Servo Drive	192.168.1.12	00-D1-00-11-41-02	6	6
12	Station 3	192.168.1.13	0A-AD-22-22-44-35	7	7
13	Station 4	192.168.1.14	00-DE-F0-13-04-01	7	
14	Station 5	192.168.1.15	0A-FF-F2-21-31-12	7	
15	PC	192.168.1.16	00-D8-88-88-88-81		8
16	Data Server	192.168.1.17	00-B0-B0-53-45-75		

FIGURE C–3   Switch lookup table contains MAC and port addresses (within dotted lines)

EXAMPLE 3: SWITCHED NETWORK WITH REMOTE CONNECTIONS

This example is a small LAN employed in an environmental control system handling temperature-related I/O and control data. While not a large application in quantity of nodes, this example is an excellent case of how an iE network functions as an integrating infrastructure for a control system and information applications for recording, analyzing, and monitoring the process locally and remotely.

In this case, the integrator is more concerned with the proper operation of control devices and applications more than with network design and performance per se. A single switch handles all the logical and physical network segmentation required. The integrator takes advantage of the vendor's use of XML-based application-level protocols to create a solution that satisfies both control and information application issues.

Fault Tolerant, Automated DNA Storage[2]

Genomics Collaborative, in Cambridge, Mass., maintains a global repository of DNA, sera and tissue samples. This repository contains samples from across the globe currently housing samples from more than 100,000 patients.

The samples are stored within a refrigerated, automated storage. The storage can house more than 1.2 million samples. The robot manufactured by RTS Thurnal Ltd., Manchester, UK, can store, retrieve, or pick 3,000 samples per day. An automated picking station acts as complex sorters and sample managers utilizing bar-code readers and tray transfer mechanisms. The robot scans the bar codes and then selects, removes, and places the sample tubes on one of several predefined locations. Automating the process ensures accurate and quick turnaround for researchers.

STORAGE

It is critical that the samples be stored in a stable, refrigerated environment (Figure C–4). CES provided the fault tolerant storage, refrigeration, and instrumentation system. The samples are stored within two repository suites. One suite consists of a 4° C picking and storage area. A duplicate suite consists of a 4° C picking room and a –20° C storage room.

Each suite is maintained at a fully redundant refrigeration system. Samples are stored with a very uniform and constant temperature environment. Backup UPS and a gas driven generator maintain power to the robotics, refrigeration system, and the Opto 22 based Ethernet control system.

INSTRUMENTATION

CES designed and installed an Opto 22 Ethernet based LCM4 controller with Opto 22 Snap I/O modules. The instrumentation system modulates the refrigeration loads to maintain a +/– 1° C temperature within the storage suites. The Ethernet based controller monitors suction and discharge pressures, room temperature and relative humidity, compressor operation, equipment amperage draw and the state of commercial power. The Opto 22 instrumentation system resides on the client's Ethernet PC network. The control system and its associated input/output devices appear just as another IP address on the network. The controller and local I/O modules are connected via a 10/100-BaseT Ethernet switch (Figure C–5).

A local PC supports Opto 22's Opto Display graphical interface. The HMI provides an animated review of the critical temperature and humidity data and the mechanical systems

---

[2]Text and graphic content courtesy of Controlled Environment Structures, Inc., Mansfield, Mass., www.CESweb.com.

FIGURE C–4   PC control and monitoring station
(Courtesy of Controlled Environment Structures, Inc.)

performance. The HMI gathers the data and generates a historical record. Given the Ethernet connectivity, the scientists are able to review the storage's operation from any PC of the network.

Password security and an electronic audit trail of user activity ensure the system's security.

COMMUNICATIONS
Due to the critical nature of monitoring the repository temperature data, the fault tolerant design carries over to the PC network. The local PC that gathers the data for a historical record must stay on line. Therefore the communications between the Windows NT based HMI and the Opto 22 controller are actively monitored. If for any reason communications were to stop, another PC of the network would automatically launch a redundant HMI to continue the data-gathering task.

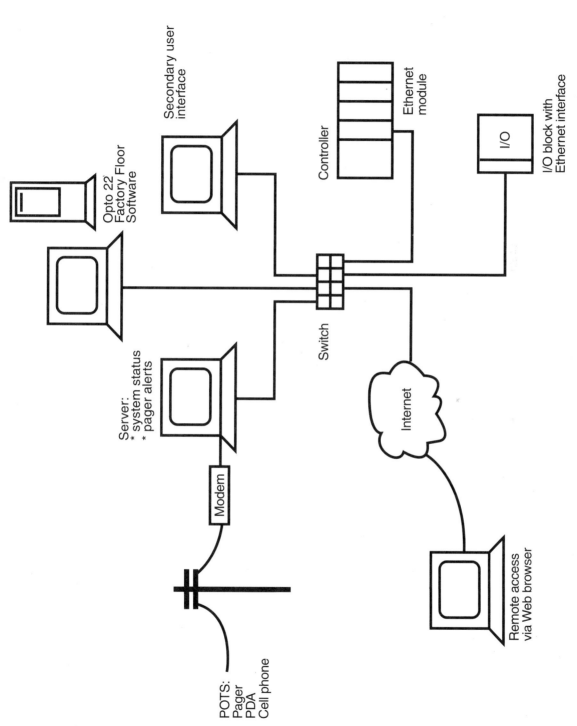

FIGURE C–5   Network architecture for an environmental control system

(Courtesy Controlled Environment Structures, Inc.)

285

ALARMING—EMAIL/PAGER ALERTS

Opto's Ethernet controller directly dispatches descriptive email alerts upon a mechanical fault or temperature issue. Emails are routed to pagers, cell phones, and/or PDA's. The client is able to gain full off-site access to the system through their VPN.

EXAMPLE 4: TIERED NETWORK

This example is much larger in scale, because it is a WAN encompassing an entire manufacturing complex. Due to size, controllers and I/O blocks are not shown in the topology diagram, but rather hubs, switches, and other gateway devices. At this scale, network reliability becomes a major issue, along with the proper selection of network components. Consequently, the network designer relies on the vendor's ability to provide redundancy by employing switches that can detect a break in the backbone. Although the topology appears to be a ring, the dotted line indicates that the switch doesn't allow a loop to be completed in normal conditions. Pass-through communication is only allowed if a break is detected, in which case, of course, a loop still isn't formed.

Automotive Assembly Plant, Volkswagen[3]

Automotive manufacture makes rigorous demands of networking technology. VW's Mosel plant is built on a Gigabit Ethernet backbone, integrating both administrative and production areas.

Volkswagen Saxony in Mosel near Zwickau, Germany, manufactures Passat limousines and the Golf IV. The factory employs nearly 5,900 workers involved in the manufacture of in excess of 1,000 cars a day. In July 1999, the millionth VW rolled off the assembly line out of a total of 268,000 vehicles for the year. The communication network used throughout the factory is based on Hirschmann technology, using all three main automation product lines, from star couplers through workgroups and the GLS (Gigabit LAN switch) backbone switch to industrial DIN rail mounted Ethernet products.

Hirschmann components have been in use at VW Saxony in Mosel since 1991, and have been used in the VW corporate group for several years before that. Hence, the network has evolved using Hirschmann equipment. Two years ago, it was decided not to use ATM technology and remain with Ethernet. The perceived relative ease of controlling Ethernet technology also played an important role in the decision to go with Gigabit Ethernet in the backbone. The growing requirements of the VW Saxony network in Mosel have been met by the expansion of the Hirschmann product range.

Assembly of the Passat and Golf is to individual customer specification and therefore individual data packets regarding special requirements, such as engine size and color, are available over the network.

---

[3]Text reprinted by permission of *Industrial Networking and Open Control* magazine: www.industrialnetworking.co.uk. Graphic content courtesy of Hirschmann Electronics, Inc.

Network

Protection from failure, especially in manufacturing, and the growth of the network play a key role in network planning today. As quantity and quality of data has increased, network management has taken on greater significance and switching bandwidth and transmission speed are now factors of moderate importance.

Interference. The data network at Mosel connects around 1,200 users over 1.8 sq. km (Figure C–6). The total length of cable laid for the network is over 1000 km, of which nearly 70% is optical fiber to just over 30% copper cabling. The longest distances between connections on site are about 2.5 km. Therefore, mono-mode optical fiber technology has also recently been installed for Gigabit Ethernet. As well as distance, resistance to interference (for example, protection from electromagnetic interference for the optical fiber lines) is a decisive factor determining usage. Multi-mode optical fiber is also used in the structured cabling (backbone, buildings and floors).

It would be possible for an employee, if authorized, to access all areas of the network which integrates more than 50 servers (Compaq, HP, Digital, and IBM), running on Windows NT or Unix platforms. Special demands are placed on bandwidth by the database applications required for the flow of manufacturing, server applications, high-resolution graphics and Web applications such as Internet and Intranet.

All administrative areas are integrated into the network (including accounts and human resources) as well as production areas (such as vehicle assembly, the paint shop, the trim shop, and the chassis construction area which is equipped with 650 robots). Applications include file and print service; specific databases such as Oracle, DB2, and Microsoft SQL; the SAP R/2 and R/3 systems; SNA-3270 host emulation for the mainframe computer; Web-based applications; email systems, Intranet, and Internet; and online access to corporate group resources in the computer centers at Wolfsburg and Ingolstadt.

The network is very important, not only for the administrative area but also for manufacturing. For example, the test results for final vehicle assembly are integrated into the overall data network through Ethernet Rail switches and hubs from Hirschmann's Industrial Line range.

Central databases provide information about specific customer requirements for each individual vehicle. Vehicles are identified by barcode—the information is read from a control computer and the data is then transferred to the assembly phase.

After final assembly, the data of the test results go to central control and guidance computers. In addition, the system provides information about which test cycles are necessary. For example, a vehicle with a diesel engine is subjected to different tests from a petrol vehicle.

Availability. Even when planning the network for the Mosel plant, it was imperative for it to be 100% available around the clock. It also needed to be used in various areas of manufacturing such as chassis construction or the paint shop. Therefore, the active components needed to be impervious to heat, dust, and vibration and had to be able to span great distances.

VW Saxony also places additional requirements on Ethernet switches and hubs used in manufacturing so that they take up only a small amount of space and run on the plant's 24 V DC power supply. Redundancy requirements achieved greater significance than bandwidth in the Gigabit Ethernet decision for GRS.

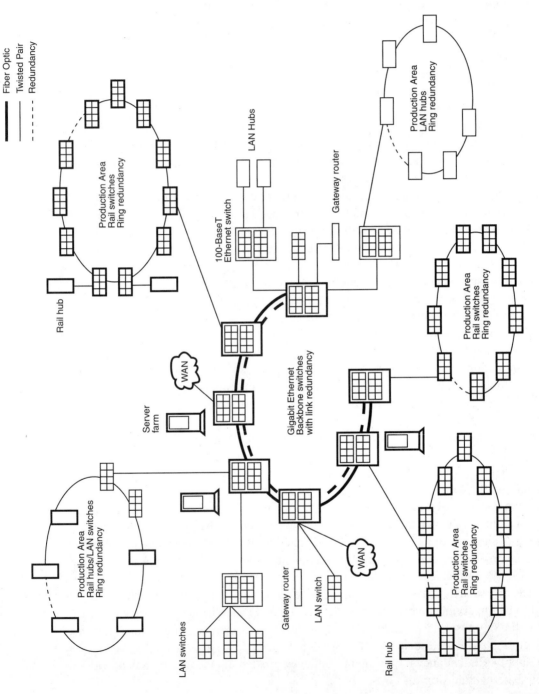

FIGURE C-6 Network architecture for car manufacturing facility
(Courtesy Hirschmann Electronics, Inc.)

The greatest demand on the network is directed at its availability. The network must be available everywhere, for both administration and manufacturing. Downtime would immediately result in a significant loss to a plant producing 1,100 cars a day. To avoid production failures, redundant networks are used in the manufacturing environment.

Redundancy. There is a good reason for the high value placed on ring redundancy in cabling. Data cables have already been damaged as a result of construction activities performed at the site. Thanks to ring redundancy, this did not result in a network crash.

"In terms of redundancy, we have made use of all the possibilities of hardware redundancy and redundant cabling structure, such as with the ASGE family, with the GRS Gigabit Routing Switch and with the Rail family," says Matthias Müller, the factory's director of organization and IS coordination. "Because of the elevated delay times, we also decided not to use the slow spanning tree.

"Redundancy is also important for us because we can no longer perform 'sneaker maintenance' as a result of the extended distances the network covers," he continues. "We have to expand the functional mechanisms that place the network independently in the position of being able to restore network functionality automatically in the event of a failure. Redundant structures in the network make it possible for us to support the network with personnel on site only between 6:00 A.M. and 6:00 P.M."

REMOTE MONITORING

Another aspect of the economic advantages provided by the network is central management. An employee can monitor the local network from the central management station and perform remote monitoring of motor manufacturing at Chemnitz or in the customer service shop at Zwickau, for example, saving an 8–10 kilometer trip. VW Saxony is further expanding the backbone with Gigabit Ethernet.

Appendix D
Acronyms

ACCO	active control connection object
ACR	attenuation-to-crosstalk ratio
ADS	autonomous decentralized system
ANSI	American National Standards Institute
API	application program interface
APO	advanced planner and optimizer
APS	advanced planning and scheduling
ARP	address resolution protocol
ARPA	Advanced Research Projects Agency
ASCII	American Standard Code for Information Interchange
ASIC	application-specific integrated circuit
ATM	asynchronous transfer mode
ATP	available to promise
AUI	attachment unit interface
AWG	American wire gauge
B2B	business-to-business
BGP	border gateway protocol
BL	blue
BOOTP	boot protocol
BR	brown
BSS	basic service set
BTO	build to order
c	speed of light (electromagnetic energy) in free space
CAD	computer aided design
CAE	computer aided engineering
CAM	computer aided manufacturing
CAN	controller area network
CAP	carrierless amplitude and phase
CATV	community antenna television
CBX	computerized branch exchange

CD	compact disk
CERT/CC	CERT Coordination Center
CIM	computer integrated manufacturing
CIP	control and information protocol
CLNP	connectionless network protocol
CMIP	common management information protocol
CMOT	CMIP over TCP/IP
CNC	computer numerical control
COM	component object module
CONS	connection-oriented network protocol
CORBA	common object request broker architecture
CRC	cyclical redundancy check
CRM	customer relationship management
CSMA/CA	carrier sense, multiple access, with collision avoidance
CSMA/CD	carrier sense, multiple access, with collision detection
DAQ	data acquisition system
DARP	dynamic address resolution protocol
DARPA	Defense Advanced Research Projects Agency
dB	decibel
dBm	decibel referenced to a milliwatt
DBMS	database management system
DCOM	distributed component object model
DCS	distributed control system
DDL	device description language
DDP	datagram delivery protocol
DES	data encryptional standard
DHCP	dynamic host configuration protocol
DIX	Digital, Intel, Xerox
DNS	domain name server
DQDB	distributed queue, dual bus
DSL	digital subscriber line
DSSS	direct sequence spread spectrum
DTE	data terminal equipment
EAI	enterprise application integration
EBCDIC	Extended Binary Coded Decimal Information Code
EDI	electronic data exchange
EGP	exterior gateway protocol
EIA	Electronic Industries Alliance
ELFEXT	equal-level far-end crosstalk
EMD	equilibrium mode distribution
EMI	electromagnetic interference

ERP	enterprise resource planning
ESS	extended service set
FAS	fieldbus access layer
FCC	Federal Communications Commission
FDDI	fiber distributed data interface
FFB	flexible function block
FHSS	frequency-hopping spread spectrum
FIP	fieldbus access sublayer
FL	factory link
FTAM	file transfer access and management
FTP	file transfer protocol
FTP	foil twisted pair
G	green
Gb/s	gigabit per second
GGP	gateway-to-gateway protocol
GHz	gigahertz
GOSIP	government open-system interconnection profile
HMI	human machine interface
HSE	FOUNDATION Fieldbus high-speed Ethernet
HTML	hypertext markup language
HTTP	hypertext transfer protocol
HVAC	heating, ventilation, and air conditioning
IANA	Internet Assigned Numbers Authority
IARP	inverse address resolution protocol
IC	intermediate cross-connect
ICANN	Internet Corporation for Assigned Names and Numbers
ICMP	internet control message protocol
ICQ	"I Seek You"
IDA	interface for distributed automation
IDC	insulation displacement contact/connector
IDF	intermediate distribution frame
IDL	interface definition language
IDRP	interdomain routing protocol
IEC	International Electrotechnical Committee
IEEE	Institute of Electrical and Electronic Engineers
IETF	Internet Engineering Task Force
IGRP	interior gateway routing protocol
I/O	input/output
IP	internet protocol
IPC	interprocess communication
IPSec	Internet protocol security

IPv4	Internet protocol version 4
IPX	Internet packet exchange
ISA	Internet Security Alliance
ISM	instrument, scientific, medical
ISO	International Organization for Standardization
JIT	just in time
JOP/MSTC	Japan Open Systems Promotion Group of the Manufacturing Science Technology Center
kHz	kilohertz
km	kilometer
KPI	key performance indicators
L2TP	layer 2 tunneling protocol
LAN	local area network
LED	light-emitting diode
LLC	logical link control
LSE	LAN redundancy entity
MAC	media access control
MAP	manufacturing application protocols
Mb/s	megabit per second
MC	main cross-connect
MC	microprocessor-based controller
MDF	main distribution frame
MES	manufacturing execution systems
MHz	megahertz
MMI	man machine interface
MMS	manufacturing message service
MRP	material requirements planning
MRPII	manufacturing resource planning
MTO	make to order
MTU	maximum transmission unit
NA	numerical aperture
NAT	network address translation
NDDS	network data delivery service
NDS	netware directory service
NEC	National Electrical Code
NetBEUI	netbios enhanced user interface
NetBIOS	network basic input/output system
NEXT	near-end crosstalk
NFS	network file system
NIC	network interface card
NIS	network information services

nm	nanometer
NMS	network management system, network management station
NRZ	nonreturn to zero
NRZI	nonreturn to zero inverted
O	orange
OD	object dictionary
ODVA	Open Device Net Vendor Association
OLE	object linking and embedding
OPC	OLE for process control
OPC DX	OPC data exchange
ORB	object request broker
ORPC	object remote procedure call
OSI	open systems interconnect
OSI/RM	open systems interconnect reference model
OSPF	open shortest path first
PAN	personal area network
PABX	private automatic branch exchange
PARC	Palo Alto Research Center
PBX	private branch exchange
PC	personal computer
PCI	protocol control information
PDM	product data management
PDU	protocol data unit
PHY	physical (layer)
PING	packet internet groper
PLC	programmable logic controller
PMAP	port mapper
PMD	Physical Media Dependent
PO	primary output
PPTP	point-to-point tunneling protocol
PSE	power-sourcing equipment
PSELFEXT	power sum equal-level far-end crosstalk
PSFEXT	power sum far-end crosstalk
PSI	pounds per square inch
PSNEXT	power sum near-end crosstalk
PV	process variable
QoS	quality of service
RARP	reverse address resolution protocol
RF	radio frequency
RFC	requests for comments
RIP	routing information protocol

RJ	registered jack
RLOGIN	remote log in
RMI	Java remote method invocation
RMON	remote monitoring
ROI	return on investment
RPC	remote procedure call
RSA	Rivest, Shamir, Adleman
RSVP	resource reservation protocol
RTMP	routing table maintenance protocol
RTPS	real-time publish-subscribe
RTS	request to send
RTU	remote terminal unit
RWhois	remote whois
RZ	return to zero
SCADA	supervisory control and data acquisition
SCM	supply chain management
scTP	screened twisted pair
SCU	service data unit
SFF	small form factor
SMF-PMD	Single-Mode Fiber Physical Media Dependent
SMTP	simple mail transfer protocol
SNMP	Simple Network Management Protocol
SOAP	simple object access protocol
SONET	synchronous optical network
SPX	sequenced packet exchange
SRL	structural return loss
STP	shielded twisted-pair cable
sUTP	screened unshielded twisted-pair cable
Tbps	terabit per second
TC	telecommunications cross-connect
TCP	transmission control protocol
Telnet	telecommunications network protocol
TFTP	trivial file transfer protocol
TIA	Telecommunications Industry Association
TP-PMD	Twisted Pair Physical Media Dependent
TTL	time to live
TTL	transistor-transistor logic
UCA	utility definition language
UDDI	universal description, discovery and integration
UDP	UNIX, user, or unreliable datagram protocol
USOC	Universal service order code

UTP	unshielded twisted-pair cable
V	volt
VAS	variable access service
VCSEL	vertical-cavity surface-emitting laser
VLAN	virtual area network
VMI	vendor-managed inventory
VPN	virtual private network
W	white
W3C	World Wide Web Consortium
WAN	wide area network
WAS	warehouse automation system
WIP	work in process
XDR	exchange data representative protocol
XML	extensible markup language
λ	wavelength
μm	micrometer

Appendix **E**
Glossary

1000BASE-LX Gigabit Ethernet over fiber with a long wavelength (1300-nm) source.

1000BASE-SX Gigabit Ethernet over fiber with a short wavelength (850-nm) source.

1000BASE-TX Gigabit Ethernet over category 5 unshielded twisted-pair cable at 1000 Gb/s.

100BASE-FX Fast Ethernet over fiber-optic cable at 100 Mb/s.

100BASE-T4 Fast Ethernet over category 3 unshielded twisted-pair cable at 100 Mb/s and using four cable pairs.

100BASE-TX Fast Ethernet over category 5 unshielded twisted-pair cable at 100 Mb/s and using two cable pairs.

10BASE-2 Ethernet over thin coaxial cable at 10 Mb/s

10BASE-5 Ethernet over thick coaxial cable at 10 Mb/s.

10BASE-F Ethernet over fiber-optic cable at 10 Mb/s.

10BASE-T Ethernet over unshielded twisted-pair cable at 10 Mb/s.

access control list A table maintained by Outlook Exchange servers containing source, destination, and port number ranges used in valid packets.

access method The rules by which a network device gains the right to transmit on the network. Common methods include carrier sense/multiple access, with collision detection, and token passing.

advanced planning and scheduling (APS) Software that remedies ERP's inability to handle real-time trend data.

aliasing Differences between the shape of an analog signal and its digital reconstruction.

analog signal An electronic pulse that varies continuously. The fluctuating voltage or frequency of an electric current represents by *analogy* the flux of a control variable, such as the rise and fall of temperature or pressure.

architecture The overall structure of logical interrelationships between physical and/or conceptual components.

asynchronous Events that occur without a time pattern. Asynchronous transmissions employ start/stop codes to signal the beginning and end of a character instead of the end of time intervals.

attenuation A decrease in power from one point to another.

attenuation-to-crosstalk ratio (ACR) The difference between the power on a received pair to the crosstalk on the same pair, giving an indication of the margin between the noise level and signal level. The ACR typically decreases with increasing frequency.

available to promise (ATP)   A strategy that enables businesses to manage fulfilling customers orders based on a knowledge of inventory availability, production capacity, and delivery requirements. A by-product of this approach is the ability to provide customers with an accurate delivery date.

backbone   The main circuit from switch to switch that carries the traffic from a number of attached ("dropped") lines.

backbone cabling   Cable running between buildings or between telecommunications closets. Vertical cable. In star networks, the backbone cable interconnects hubs, switches, and similar devices, as opposed to cables running between hub and station. In bus networks, the bus cable.

balanced transmission   A mode of signal transmission in which each conductor carries the signal of equal magnitude but opposite polarity. A 5-volt signal, for example, appears as +2.5 volts on one conductor and −2.5 volts on the other.

balun   An impedance-matching device that also provides conversion between balanced and unbalanced modes of transmission. The term derives from "balanced/unbalanced."

bandwidth   The frequency range that can be used by a signal, expressed in units of hertz (Hz).

binary digits   Machine language represented by the numbers 0 and 1

binding   The action of associating an IP address and an interface's media access control (MAC) number.

blocking   The technique of halting an application until a reply has been received; a characteristic of **tightly coupled** connections.

bridge   A network device that interconnects two networks allowing all stations to share the combined network, while isolating transmissions on one segment from colliding with transmissions on the other. Using a bridge makes it possible to segment a network into separate collision domains. A bridge reads the MAC address determined by the data-link (OSI layer 2) protocol and will only forward the packet if the address is unknown.

broadcast domain   The portion of a network where a transmission from one station can be seen by all the other stations.

build to order (BTO)   A production strategy where the production process is triggered by the customer's order.

bus   The arrangement of a single cable or multiple cables in a single group that provides a common connection to many devices.

business-to-business (B2B)   Network facilitated transactions of supplies and commodities that usually occur in "real time," meaning in this case, a time period that satisfies the user.

byte   A unit of code comprised of eight binary digits.

carrier sense multiple access with collision detection (CSMA/CD)   A network access method used by Ethernet in which a station listens for traffic before transmitting. If two stations transmit simultaneously, a collision is detected and both stations wait a brief time before attempting to transmit again. CSMA/CA (collision avoidance) broadcasts a request-to-send frame to ensure a clear wire before message transmission and is used for wireless Ethernet.

cell   A group of workstations, machines, or equipment organized to perform a complete sequence of operations.

cellular manufacturing   A manufacturing approach that organizes resources in a single group or cell to produce a family of parts. The goal is to reduce lead time and inventory by employing pull scheduling, visible controls, quality at the source, and improved team work. The concentration of resources requires a high level of communication, which makes it difficult for a traditional fieldbus network to carry the megabyte-sized data and messages required by the cell.

circular network effect   The phenomenon whereby technology that increases in information flow creates the need for more information technology.

cladding   The outer concentric layer of an optical fiber, with a lower refractive index so that light is guided through the core by total internal reflection.

client   The destination device issuing a request to the server.

closed-loop control   A system in which an output is fed back as an input to adjust the system's performance.

coaxial cable   A cable in which the center, signal-carrying conductor is concentrically centered within an outer shield and separated from it by a dielectric. The structure, from inside to outside, is center conductor, dielectric, shield, and then outer jacket.

cockpit   See *dashboard.*

collision   The signal conflict that occurs when two stations on a common Ethernet network segment broadcast simultaneously.

collision domain   A single segment of a network where collisions can occur. On a single-segment network, the broadcast domain and collision domain are one and the same. On a multisegment network, a broadcast domain can include multiple collision domains.

communications network   Premises cabling that deals with the lowest layer, the physical layer.

computer-integrated-manufacturing (CIM)   A facility that operates with minimal direct human intervention.

conduit   A raceway of circular cross-section. Also used more generally to indicate an enclosed cable pathway.

controller   A stand-alone device, such as a PLC, or a program in a PC that receives inputs from a system, performs control calculations, and generates an output used to adjust the system.

core   The central light-carrying portion of an optical fiber.

critical to quality   The key selling point.

crosstalk   The unwanted transfer of energy from one circuit to another. Crosstalk can be measured at the same (near) end or far end with respect to the source of energy.

customer-relationship management (CRM)   An approach that orients designing, managing, purchasing, production, and other aspects of the business around customers' needs. To handle the communication needs of such an approach, special software applications and systems may be required.

dashboard   The graphical interface in an ERP or enterprise portal system that gives management a complete view of the most important business activities, known as *key performance indicators (KPI).*

data acquisition system (DAQ)   An digital instrument or PC that gathers data from sensors. It is used in conjunction with amplifiers, multiplexers, and analog-to-digital converters. Often employed in process industries.

data connector   A four-position connector for 150-ohm STP and used primarily in Token Ring networks.

data historian   See *data acquisition system.*

data warehouse   A database used for analytical research rather than day-to-day business transactions.

datagram sockets.   The technique of using UDP to associate an IP address and a port number to create a connectionless method of data transport that *sends individual packets to a destination.*

DC loop resistance   The total DC resistance of a cable. For a twisted-pair cable, it includes the round-trip resistance down one wire of the pair and back up the other wire.

DCS (distributed control system)   Evolving from centralized process control computers in the 1960s, DCS handles continuous flow processes involving open-loop and analog control. Due to the nature of continuous and complex batch-process applications, DCS requires "real-time" performance in both the control system and the control network. Since these applications are associated with large chemical, water, waste-water, and utility plants, the terrain makes it advantageous for devices and controllers to be located near the operation and away from a centralized control room. A DCS is networking intensive because the control information is usually transmitted from many I/O modules over a single bus to a central control room.

dead system   A device that is not operating.

decapsulation   The process of opening a frame and processing the contents without disturbing the other frames nested within.

decibel   The logarithmic ratio of two powers, voltages, or currents, often used to indicate either gain and loss in a circuit or to compare the power in two different circuits. dB = $10 \log_{10} (P_1/P_2) = 20 \log_{10} (V_1/V_2) = 20 \log_{10} (I_1/I_2)$. Power is most often used in premises cabling.

dense wavelength-division multiplexing (DWDM)   A form of optical multiplexing in which different wavelengths (signals) are transmitted simultaneously through a fiber. The wavelengths are all in the same optical windows and can be separated by about 1 nm.

deterministic   The ability of a system to process every event as they occur.

devices   An electronic or mechanical mechanism that can send a signal and/or perform an action. In control, field devices are the physical elements that affect a controlled system and are typically located at a distance from the controller. Input devices can be switches (on/off, pushbutton, limit, proximity, etc.); output devices can be lights, motor starters, pumps, and solenoids that drive valves. Measuring devices include flow, temperature, and pressure sensors as well as more complex instrumentation.

digital signal   An electric signal that uses defined levels of changing amperage or voltage to encode binary values (0,1).

dirty signals   The digital or analog transmission of data that is corrupted by electromagnetic interference.

disintermediation   The effect of Internet-based business transactions to cut out middlemen.

dispersion   A general term for those phenomena that cause a broadening or spreading of light as it propagates through an optical fiber. The two main types are modal and material.

distributed computing, distributed control, distributed I/O   As a general concept, a distributed approach breaks a complex centralized system into modules that can perform operations at dispersed locations and transmits outputs to other devices. The distributed concept depends on inexpensive processors and pervasive networking. Thus, distributed computing involves modular program components (such as Java applets or CORBA objects) instead of a single program; distributed control involves control functions embedded in "smart" field devices instead of a single PLC or PC; distributed I/O involves I/O modules in the field instead of local I/O boards in a centralized PLC.

driver A small program that handles the input and output request between the system and the hardware.

dynamic packet filtering A technique used in routers and firewalls that uses stateful inspection which validates packets based on whether their contents were propagated in response to a client's request.

electronic data interchange (EDI) A template that allows data to be imported into any database that uses that template. Commonly used by credit card companies.

embedded control The incorporation of a microprocessor in instruments and sensors to perform control functions and network communications.

encapsulation The process of forming a sequence of digital envelopes to contain data; one envelope or frame is nested inside another.

enterprise application integration (EAI) See *middleware.*

enterprise resource planning (ERP) The broad set of computer- and network-based activities enabled by application software involving modules for managing product planning, parts purchasing, inventories, supplier interaction, customer service, order tracking, finances, and human resources. Consequently, an ERP system involves complex databases, rigorous process analysis, and employee training.

equal-level far-end crosstalk (ELFEXT) Crosstalk that is measured on the quiet line at the opposite end as the source of energy on the active line and relative to the received signal level. FEXT minus attenuation equals ELFEXT.

Ethernet A protocol originally developed by Robert Metcalfe and collaborators at Xerox and now associated with the IEEE 802.3 standard. Ethernet establishes physical and data link (OSI layers 1 and 2) features for cabling, signal encoding, MAC address header, and the overall frame for a data packet.

far-end crosstalk (FEXT) Crosstalk that is measured on the quiet line at the opposite end as the source of energy on the active line.

fieldbus A large family of bidirectional, digital communication protocols that were specially developed to overcome the physical and performance limitations of low-level serial and analog standards.

File Transfer Protocol (FTP) An application layer protocol that runs on top of TCP/IP to allow a user to directly transfer files between hosts.

firmware A program stored in a memory chip.

flow manufacturing A production method that allows one production line to produce a fixed number of products with a variable range of features within a given period of time. This method requires precise timing between workstations to maintain a coordinated production rate.

fragmentation The process in which a gateway router's IP protocol breaks the datagram in smaller parts to fit the size requirements of another type of network

full duplex Signal transmission over one connection between two devices in both directions simultaneously.

gateway A network device that converts data formats for transmission over another kind of network based on upper-layer (OSI layers 5, 6, or 7) protocols.

graded-index fiber   An optical fiber whose core has a nonuniform index of refraction. The core is composed of concentric rings of glass whose refractive indices decrease from the center axis. The purpose is to reduce modal dispersion and thereby increase fiber bandwidth. The 50/125 and 62.5/125-μm fibers have a graded index.

greenfield   A manufacturing plant that is a completely new facility.

half duplex   Signal transmission over one connection between two devices in one direction at a time.

hard real-time   A method of control that times events calculated on a predictable basis.

header   A segment of code that is added to the front of data containing information that allows the data that follows it to be properly processed.

hex   Code based on grouping of binary numbers in a four-bit group.

HMI (human-machine interface)/MMI (man-machine interface)   A display panel used with a machine controller, PLC, or PC that allows the operator to see a graphic representation of the operation for monitoring and control.

home run wiring   Cabling that is used to connect field devices to I/O blocks inside the PLC cabinet.

hop   A trip a packet makes from one node point in the network to another point.

horizontal cabling   That portion of the premises cabling that runs between the telecommunications outlet of the work area and the horizontal cross-connect in the telecommunications closet.

horizontal cross-connect   The cross-connect that connects horizontal cable to the backbone or equipment.

host   A device that can communicate with another device.

hub   A network device providing physical ports to connect the Ethernet cabling, but without the intelligence to direct transmissions in any given direction.

hypertext markup language (HTML)   Tags that are used by the Web browser to determine the size, position, and font of the data being displayed.

hypertext transfer protocol (HTTP)   The header performing application-layer functions by making it possible for HTML files to be exchanged over the Internet.

I/O module   Input/output is a general term for data entering or leaving a device. In a control system, an I/O module contains circuits to process a number of inputs received from various measuring devices into digital outputs that are transmitted to a PLC or PC. Digital modules interface with on/off sensors/actuators; analog modules use D/A and A/D converters to interface with analog instrumentation; intelligent I/O modules can provide an interface for high-speed counters, programming, motion control, or communication.

IEEE 802.3   The standards committee defining what is commonly known as Ethernet networks.

infrastructure   A foundational tool of the economy or of a business that benefits everyone, like the highway system, or information technologies, like the Internet.

insertion loss   The loss of power that results from inserting a component into a circuit. For example, a connector causes insertion loss across the interconnection (in comparison to a continuous cable with no interconnection).

instrument   A device that detects a state like a sensor, but can also observe, measure, provide data, and may even control a value.

Internet protocol (IP)   The rules providing connectionless network (OSI layer 3) functions, which include numbering packets with source and destination addresses and a time to live (TTL) code. Along with TCP, it is a standard protocol for the Internet.

interprocess communication (IPC) The use of procedures in different programs that allow them to run concurrently to exchange data and parameters.

intrinsic safety A characteristic of control wiring or devices whereby voltages and currents are kept below the levels that can ignite an atmosphere.

jitter The variation of a mistimed signal in relation to its original reference time.

jumper A cable without connectors, typically used at the cross-connect to join circuits. Similar to a patch cable (which has connectors).

just-in-time (JIT) A production system that organizes the plants resources to respond to actual demand, allowing customers to pull product out of the factory (compare with MRP). Popularly associated with reducing material costs and lead times by shifting the burden of managing inventory to the supplier.

laser A light source producing, through stimulated emissions, coherent, near-monochromatic light—an intense, narrow beam at a single frequency. Lasers in fiber optics are solid-state semiconductors.

latency The time it takes for an electronic pulse to be conveyed over a medium, from origination to destination. In networking, it is the time a switch takes to figure out where to forward a data packet.

light-emitting diode A semiconductor that spontaneously emits light when current passes through it.

loosely coupled A connection that allows a sender to communicate an arbitrary number of receivers.

manufacturing execution systems (MES) A high-level information system used in the plant for quality control, document management, plant-floor scheduling, and dispatching. An MES tracks work-in-process (WIP) by collecting data on product routing and tracking, labor, resources, and production. It also collects "live" information about setups, run times, throughput, and yields to identify bottlenecks and capacity problems. Consequently, the MES is critical to plant floor *and* enterprise communication by delivering real-time order status to the supply chain, by enabling an available-to-promise capability for the customer, and by updating ERP systems for business managers.

manufacturing resource planning II (MRPII) A system that allows purchasing, marketing, engineering, finance, and accounting to use outputs from MRP to coordinate work and achieve cost savings.

manufacturing system The method of organizing the design process, manufacturing planning and control, and production operations based on the nature of goods or services being produced. Five systems include: project organization to build one-of-a-kind products (airplanes), job shops for low volume goods (machine tools), repetitive systems for high volume production (valves), line systems using product **flow** layouts to accommodate products with many pieces (computers, cars), and continuous process systems for standard products in very high volumes (bolts, gasoline).

marshalling The process of preparing parameters for serial transmission on the local host.

material dispersion Dispersion that results from each wavelength traveling at a different speed than other wavelengths through an optical fiber. Also called *chromatic dispersion.*

material requirements planning (MRP) A system that organizes plant resources to respond to predictated demand, resulting in the manufacturer pushing products to customers. This approach is enabled by software used to handle the ordering and the bill of materials for products with hundreds of parts.

message   The information that a user wishes to transmit over a network.

method   The command that executes code in object-oriented programs.

middleware   A general term for any programming that stands between two separate applications and allows them to interoperate. In software, middleware is a program that allows one database to access another database. In communication, middleware is a messaging service that allows different applications to communicate. In enterprises, middleware is used to tie together separate applications in a major initiative known as *enterprise application integration (EAI).*

modal dispersion   Dispersion that results from the different transit lengths of different propagating modes in a multimode optical fiber.

modular jack   The equipment-mounted half of a modular interconnection.

modular plug   The cable-mounted half of a modular interconnection. While available in 4, 6, and 8 positions, the 8-position version is specified for premises cabling.

modulating   Changing the frequency of the analog signal's sine wave.

multimode fiber   An optical fiber that supports more than one propagating mode.

near-end crosstalk (NEXT)   Crosstalk that measures on the quiet line at the same end as the source of energy on the active line.

network   The communication connections between computers, workstations, and related devices. **Industrial networks** are used to convey critical control information and operational data to operators, equipment, controllers, valves, and sensors. Networks are usually classified as **local area networks** (LANs) located in a limited geographical area, such as an office or factory, and **wide area networks** (WANs), located outside the physical area of the facility but that provide communications necessary for the facility's and business's operation. Together, LANs and WANs comprise the combined business network known as an **enterprise.**

network architecture   The pattern in which nodes are wired together.

network interface card (adaptor) (NIC)   A circuit board that contains the encoding chips with a unique media access control number to enable an Ethernet signal to travel over the cabling.

node   An addressed point on the network, including switches, routers, and hosts.

object-oriented programming: A method of writing code that constructs entities that respond to events according to their own code segments, known as **methods.**

Open Standard Interconnect   A seven-layer ISO model for defining an **open standard,** which means Ethernet procedures are developed and published for public access and use, either free or for a nominal fee.

openness   A design characteristic of components made by different vendors that allows them to interoperate within a system.

optical fiber   A thin glass or plastic filament that propagates light.

optical time-domain reflectometry   A method for evaluating optical fiber based on detecting and measuring back-scattered (reflected) slight. Used to measure fiber length and attenuation, evaluate splice and connector joints, locate faults, and certify cabling systems.

OSI model   A template developed in 1978 by the International Organization for Standardization (ISO) for designing protocols. Involves seven layers—physical, data-link, network, transport, session, presentation, and application—describing functions that are necessary for successful network communication.

overhead. The quantity of code involved to perform a task, usually a pejorative.

packet A unit of code and data that contains all the necessary information to travel over the network.

packet switching The routing of digital signals by encoding packets with the addresses of the sending and receiving devices.

patch cable A cable terminated with connectors and used to perform cross-connects. Similar to a jumper cable (which has no connectors).

patch cord A patch cable.

patch panel A passive device, typically a flat plate holding feed-through connectors, to allow circuit arrangements and re-arrangements by simply plugging and unplugging patch cables. The feed-through connectors can have the same or different interfaces on either side.

pen plotters An early method of recording data that uses ink pens to mark a roll of paper.

pipe A one-way communication link that allows information to be passed from the server's application program to the client's browser.

plastic fiber An optical fiber made of plastic rather than glass.

PLC A controller with a microprocessor, memory, and programming used to control large numbers of discrete elements using very fast I/O scan times. PLCs are used with other computers and applications, such as human-machine interface panels, and with data historians for data storage and evaluation.

plenum The air-handling space between walls, under structural floors, and above drop ceilings used to circulate and otherwise handle air in a building. Such spaces are considered plenums only if they are used for air handling.

plenum cable A flame- and smoke-retardant cable that can be run in plenums without being enclosed in a conduit.

point of control/data point A value representing the state of a controlled variable, such as on/off or a temperature reading, transmitted as bits; or a measurement, usually transmitted as bytes.

port The end point of the connection where the data leaves the packet and enters the application.

portal A data connection framework employing a server, database, tagging, and network and Internet connections that allows access to information contained in manufacturing, enterprise, and other systems.

power sum In NEXT and FEXT testing, a measurement of crosstalk that takes into account the sum of crosstalk from all other conductors coupling noise onto a quiet line.

product data management (PDM) The use of a data repository to manage blueprints, specifications, recipes, and drawings normally stored on paper.

production strategy The method of meeting customer demand by organizing plant and enterprise resources to reconcile manufacturing lead times and customer lead times. Five strategies are custom manufacturing, make to order, assemble to order, available to promise, and build to stock.

protocol control information The code in a header that allows the contents to be properly processed.

protocol data unit (PDU) The technical name for a data packet.

protocols A set of rules to establish a communication exchange and to create a basic digital container or envelope, known as a **frame,** to enclose other codes and data necessary for network communication.

pull manufacturing   An approach to manufacturing where the customer's order triggers production.

pull scheduling   A method of production timing used with JIT strategies that authorizes the release of work based on cell status.

push   The technique of sending data to a designated device without a specific request.

raceway   Any channel designed to hold cables.

remote diagnostics   The use of network and wireless communication technologies to access data from geographically distant industrial systems and components for service and maintenance.

repeater   A device that receives, amplifies (and sometimes reshapes), and retransmits a signal. It is used to boost signal levels and extends the distance a signal can be transmitted. It can physically extend the distance of a LAN or connect two LAN segments.

report by exception   The control programming technique of transmitting **data** only when there is a change in status.

requests for comments (RFCs)   Technical notes used in Internet development that discuss relevant aspects of computer networking, including procedures, programs, and concepts.

ring network   A network topology in which terminals are connected in a point-to-point serial fashion in an unbroken circular configuration. Many logical rings are wired as a star for greater reliability.

riser   A backbone cable running vertically between floors.

router   A network device that directs packet to the proper network based on the packet's IP address determined by network (OSI layer 3) protocols.

sampling   A measurement taken at regular time intervals.

SC connector   A fiber-optic connector having a 2.5-mm ferrule, a push-pull latching mechanism, and the ability to be snapped together to form duplex and multifiber connectors.

SCADA (Supervisory Control and Data Acquisition)   Evolving from PLC operator interfaces, SCADA is a software-based system used for process monitoring and data acquisition. It uses software-based logic to perform PLC functions on a PC. Software can add production management capabilities to allow supervisory control systems to perform at an MES level.

scan time   The time required for a PLC to completely execute a program once, including an update of I/O values.

sensor/transducer/transmitter   A transducer is a device that converts signals from one physical form (air pressure) to another (electrical current). In a broad sense, a sensor is a transducer that uses the state of a physical phenomenon as an input and returns a quantitative measure as an output. A sensor functions as a transmitter when the measurement is transmitted as a digital signal to a PLC or PC.

sensor bus   A network that carries low-level data traffic, such as one-bit values for on/off (contact) I/Os, simple transducers, and measurements.

server   The source device replying to a request made by the **client.**

services   Specific functions provided by various protocols in order to form and transmit packets to create a communication session.

seven-layer model   See *OSI model.*

shielded twisted-pair (STP) cable   A type of twisted pair cable in which the pairs are enclosed in an outer braided shield, although individual pairs are usually also shielded.

simple object access protocol (SOAP)   A specification on encoding HTTP headers and XML files so an application on one host can call an application on another to pass it information. The specification also covers how the application should return a response.

single-mode fiber   An optical fiber that supports only one mode of light propagation above the cut-off wavelength. Single-mode fibers have extremely high bandwidths and allow very long transmission distances.

smart device   A sensor with the embedded intelligence to identify itself on the network.

SMTP (Simple Mail Transfer Protocol)   An application layer protocol that runs on top of TCP/IP. Supports basic e-mail delivery services.

sniffers   A tool used by network troubleshooters and hackers. A sniffer application installed on a PC sets the NIC card into **promiscuous mode,** enabling it to accept all the broadcast traffic within the system's collision domain. The packet's binary signal is then decoded into hexadecimal and common language forms, allowing the contents in the headers and the payload to the packet to be analyzed.

socket   Created by a separate computer-program that reads and writes the protocol control information into the proper header payload format. Sockets are used whenever a client or server program application requests a network service.

Sonet   Synchronous optical network, an international standard for fiber-optic-based digital telephony.

spoofing   The hacker technique of redirecting network traffic to another device before it reaches its intended destination.

star network   A network in which all stations are connected through a single point. Star configurations tend to be reliable.

stateful inspection   See *dynamic packet filtering.*

stateless connection   The condition where any packet can be delivered anytime regardless of packet order.

static packet filtering   A technique used in routers and firewalls of looking at the source, destination, and port fields in the datagram header and either forwarding or blocking the packet.

station   A point on a network where input or output occurs.

steady-state   The minimum speed a closed-loop system must maintain to avoid going out of control.

step-index fiber   An optical fiber, either multimode or single mode, in which the core's refractive index is uniform throughout so that a sharp demarcation or step occurs at the core-to-cladding interface. Step-index multimode fibers typically have lower bandwidths than graded-index multimode fibers.

stream sockets   The technique of using TCP to associate an IP address and a port number to create a connection-oriented method of data transport that handles a continuous number of packets.

strength member   That part of the fiber-optic cable that increases the cable's tensile strength and serves as a load-bearing component. Usually made of Kevlar aramid yarn, fiberglass filaments, or steel strands.

strong password   A phrase typed into a system to gain entrance that combines ASCII upper- and lower-case characters, plus numbers and symbols, at least eight characters long.

supply chain execution   The arrangement and delivery of finished goods to enable maximum customer response.

supply chain management (SCM)   Applications depending on up-to-the-minute information.

switch   A network device that directs a packet to the station or stations for which it is intended. A switch is a multiport bridge that reads the MAC address determined by the data-link (OSI layer 2) protocol. An IP switch functions as a router by also reading the packet's IP address.

symbol   The smallest signal transition that can be encoded.

synchronous events   Those events occurring at a regular time interval. Synchronous transmissions employ a clock to signal the beginning and end of a character instead of start and stop codes.

telegram   The fieldbus equivalent of an IP datagram.

Telnet   An application layer protocol that runs on top of TCP/IP: to allows the user's device to operate as a terminal to run a remote host.

terminator   A device used to absorb the signal at the end of a fieldbus cable.

thicknet   A popular term for an Ethernet 10BASE-5 network.

thinnet   A popular term for an Ethernet 10BASE-2 network.

threads   Lightweight processes that provide an independent execution path to keep an application running.

three-tier architecture   The network design used in some traditional industrial networks in which a bus connects sensors and actuators at the bottom level; another bus connects intelligent devices and controllers at the mid-level; and yet another connects PLCs, DCS, SCADA, and PCs at the high control and supervisory level.

tightly coupled   A connection that allows a sender to communicate only with one receiver.

token passing   A network access method in which a station must wait to receive a special token frame before transmitting.

topology   The pattern in which a network is physically structured, such as a star, ring, or bus.

Tower of Babel effect   The creation of different languages and protocols based on the goals, needs, and interests of different groups.

transmission control protocol (TCP)   The rules providing connection-oriented transport (OSI layer 4) functions that include establishing and maintaining a connection between hosts, sequencing packets, and retransmission if a packet is lost. Along with IP, it is a standard protocol for the Internet.

twisted-pair cable   A cable having the conductors of an individual pair twisted around one another to increase noise immunity. Twisted-pair cables operate in the balanced mode.

velocity of propagation   The speed of electromagnetic energy in a medium (including copper and optical cables) in relationship to its speed in free space. Usually given in terms of a percentage. Test devices use velocity of propagation to measure a signal's transit time and thereby calculate the cable's length.

vertical-cavity surface-emitting laser   A low-cost laser that emits light from its surface rather than from its edge and well suited to operation with multimode fibers.

virtual local area network (VLAN)   A local area network connected on a logical rather than geographical basis, such as user type. A VLAN controller is used to add or change users.

virtual private network (VPN)   The technique of encrypting packets before transmission over a public network, then decrypting them at the receiving end.

voltage mode signaling   Using a low current and bipolar voltage, which means voltage swings between positive and negative.

wiring closet   A general term for any room housing equipment or cross-connects.

work area   That area of the premises cabling where users are located; the area from the communications outlet to the equipment connected to the premises cabling. Loosely, an office, cubicle, and so forth.

work in progress (process) (WIP)   The summary of work order amounts (lots) involving good parts, failed parts, and scrap from each process steps. WIP data is derived from real-time data that are supplied to the ERP system's APS module to optimize the actual execution of manufacturing (MES) operations—which is critical in build-to-order production strategies.

workstation   An area where a single operator monitors and responds to one or more conditions. It may also refer to the HMI or PC used in the area.

Appendix F
Further Resources

What follows are information resources on the subjects covered in this book. Of primary importance are the standards developed by various organizations for the protocols used in Ethernet and the layers that comprise iE. These standards are available from:

802.3:

www.ieee.org
IEEPublications Office
10662 Los Vaqueros Circle
P. O. Box 3014
Los Alamitos, CA 90720-1264
Publications Orders:
800 272 6657

TCP/IP:

www.ietf.org
Support organization:
Internet Society
1775 Wiehle Avenue, Suite 102, Reston, VA 20190, USA
703-326-9880

HTML/XML/SOAP:

www.w3.org
c/o Janet Daly, Head of Communications
Massachusetts Institute of Technology
Laboratory for Computer Science
200 Technology Square
Cambridge, MA 02139
USA
617-253-5884

ADS-net
www.mstc.or.jp/jop

Ethernet/IP
www.odva.org

Fieldbus HSE
www.fieldbus.org

iDA-RTPS
www.ida-group.org

Interbus
www.ibsclub.com

MMS Ethernet
www.nettedautomation.com

Modbus TCP
www.modbus.org

NET for Manufacturing (DCOM)
www.microsoft.com/technet/itsolutions

OPC DX
www.opcfoundation.org

PROFInet
www.profibus.com

BOOKS

Bartnikas, R. & Srivastava, K.D., eds. *Power and Communication Cables: Theory and Applications.* New York: IEE Press, 2000.

Benedetto, Sergio, Biglieri, Ezio, & Castellani, Valentino. *Digital Transmission Theory.* Englewood Cliffs, NJ: Prentice Hall, 1987.

Brandimarte, Paolo and Villa, Agostio, eds. *Modeling Manufacturing Systems: From Aggregate Planning to Real-Time Control.* Berlin, Germany, 1999.

Chiquoine, Walter A. & Hist, Elizabeth R. *The Connectivity Management Handbook.* Wilmington, DE: Transport Management Group Inc., 1995.

Dorf, Richard C. *Modern Control Systems.* Upper Saddle River, NJ: Prentice Hall, 2001.

Goldman, Steven L., et al. *Agile Competitors and Virtual Organizations: Strategies for Enriching the Customer.* New York: Van Nostrand Reinhold, 1995.

Goranson, H. T. *The Agile Virtual Enterprise: Cases, Metrics, Tools.* Westport, CT: Quorum Books, 1999.

Gralla, Preston. *How the Internet Works, Millennium Edition,* Indianapolis, IN: QUE, 1999.

Hannam, Roger. *Computer Integrated Manufacturing: From Concepts to Realisation.* Harlow, Essex, England: Addison Wesley Longman, 1997.

Kamen, Edward W. *Introduction to Industrial Controls and Manufacturing.* San Diego, CA: Academic Press, 1999.

Katebi, Reza. *Control and Instrumentation for Wastewater Treatment Plants.* London, England: Springer-Verlag London Limited, 1999.

Liautaud, Bernard with H. Hammond, Mark. *e-Business Intelligence: Turning Information into Knowledge into Profit.* New York: McGraw-Hill, 2001.

Marshall, Perry S. *Industrial Ethernet: A Pocket Guide.* Research Triangle Park, NC: ISA—The Instrumentation, Systems, and Automation Society, 2002.

Martin, James, Kavanaugh, Kathleen, & Leben, Joe. *Local Area Networks: Architectures and Implementations,* 2nd ed. Englewood Cliffs, NJ: PTR Prentice Hall, 1994.

O'Leary, Daniel Edmund. *Enterprise Resource Planning Systems: Systems, Life Cycle, Electronic Commerce, and Risk.* Cambridge, UK: Cambridge University Press, 2000.

Rehg, James A. and Kraebber, Henry W. *Computer-Integrated Manufacturing, 2nd ed.* Upper Saddle River, NJ: Prentice-Hall, Inc., 2001.

Rowe, Alan J. and Davis, Sue Anne. *Intelligent Information Systems: Meeting the Challenge of the Knowledge Era.* Westport, CT: Quorum Books, 1996.

Svacina, Bob. *Understanding Device Level Buses: A Tutorial.* Minneapolis, MN. InterlinkBT, LLC, 1998.

Scambray, Joel, McClure, Stuart, and Kurtz, George. *Hacking Exposed: Network Security Secrets and Solutions,* Berkeley, CA: Osborne/McGraw-Hill, 2001.

Seifert, Rich, *The Switch Book: The Complete Guide to LAN Switching Technology.* New York: John Wiley & Sons, 2000.

Spurgeon, Charles E. *Ethernet: The Definitive Guide.* Cambridge, MA: O'Reilly and Associates, 2000.

Stallings, William. *Data and Computer Communications,* 4th ed. New York: Maxwell Macmillan International, 1994.

Sterling, Jr., Donald J. & Baxter, Les. *Premises Cabling,* 2nd ed. Albany, NY: Delmar Publishers, 2000.

Stevens, W. Richard. *TCP/IP Illustrated, Volume 1: The Protocols.* Boston, MA: Addison-Wesley, 1994.

Svrcek, William Y., Mahoney, Donald P., and Young, Brent R. *A Real-Time Approach to Process Control.* New York: John Wiley & Sons, 2000.

Tanenbaum, Andrew S. *Computer Networks,* 3rd ed. Upper Saddle River, NJ: Prentice Hall, 1996.

Taylor, Ed. *The Network Architecture Design Handbook: Data, Voice, Multimedia, Intranet, and Hybrid Networks.* New York: McGraw-Hill, 1997.

PERIODICALS

George, Bill, ed., *The Industrial Ethernet Book,* GGH Marketing Communications, Titchfield, Hauts, Great Britain. (http://ethernet.for-industry.com)

Kvo, G.S., ed., IEEE Communications Journal, The Institute of Electrical and Electronics Engineers, Inc., New York, NY. (www.coumsoc.org/~ci)

Hale, Gregory, ed., *INTECH,* ISA-The Instrumentation, Systems, and Automation Society, Research Triangle Park, NC. (www.isa.org)

Appendix G
Answers to Odd-Numbered Questions

Chapter 1

1. HTTP.
3. IPX/SPX.
5. The logical end point of a connection where the application layer process accesses TCP services.
7. It is used to confirm packet delivery and reassembling packets into the original message.
9. ARP.
11. 26 years.

Chapter 2

1. Sensor, controller, actuator, gauge.
3. Speed, data storage, flexibility.
5. The control system must be able to adjust variables to actual events, not after the fact.
7. Ironically, a DCS is located in a centralized control room.
9. One-to-one, all-to-one, one-to-many, one-to-any.

Chapter 3

1. Removes the informational friction caused by scarce, slow, and low-quality data.
3. To increase the information available to senior management, to support e-business initiatives, to adapt to the increasing rate of change.
5. Cellular manufacturing.
7. APS—Advanced Planning and Scheduling.
9. When using iE for remote monitoring and control.

Chapter 4

1. Xerox, Intel, Digital Equipment Corporation.
3. 10 Mb/s.
5. 46, 1500.
7. Carrier sense, multiple access, with collision detection (CSMA/CD).
9. A frame operated on the data-link layer (layer 2). A packet works on the network layer (layer 3).

Chapter 5

1. 10 Mb/s, 100 Mb/s.
3. A hub is a shared-media device in which traffic flows to each attached device. A switch is a switched-media device in which traffic flows only to those stations addressed.
5. MAC address.
7. Greater reliability in the event of a cable or device failure.
9. A switched-media network using full-duplex transmission.

Chapter 6

1. Monitoring and control.
3. Polling consumes bandwidth.
5. Security.
7. Quality of service is a means to guarantee a specific level of bandwidth.
9. Midspan and DTE.

Chapter 7

1. NEXT is measured at the near end; FEXT is measured at the far end (with reference to the signal source).
3. Category 3 = 16 MHz; Category 5 = 100 MHz; Category 6 = 250 MHz.
5. Category 5i has a more rugged jacket and a wider temperature rating.
7. The core carries the light; the cladding helps guide it.
9. Multimode fiber.

Chapter 8

1. Wireless Ethernet uses a different access method from standard Ethernet. The 802.3 and 802.11 committees are based on an access method.
3. Piconet.
5. An ad hoc network does not use access points. An infrastructure network uses access points.
7. Transmission speed generally decreases as distances increase.
9. Bluetooth uses a master/slave arrangement instead of the CSMA/CA used by wireless Ethernet.

Chapter 9

1. Unicast, broadcast, and multicast.
3. The exchange of data and parameters between programs on different devices.
5. Socket programming.
7. Class D.
9. Packets can be delivered in any order.
11. A feature of TCP that causes an application to pause until it receives the requested reply.
13. Yes.
15. Data parameters, not large-sized messages.
17. No, because ports are dynamically assigned, it creates problems with firewalls.
19. To take advantage their ability to minimize network traffic by allowing one message to be delivered to a designated group of devices.

Chapter 10

1. How the protocol balances the trade-off between speed and reliability.
3. It simplifies programming.
5. It reduces network load by preventing multicast packets from being copied to every station on the WAN.
7. Port 21.
9. CIP—Control and Information Protocol.
11. It promises real-time performance at speeds allowing precise drive synchronization over the Ethernet.

Chapter 11

1. They have more potential points of access to the outside world.
3. Static packet filtering simply looks at the source, destination, and port fields in the IP header. Dynamic packet filtering uses stateful inspection to the packet's contents to ensure the packet was propagated in response to a client's request.
5. Closes ports as potential points of intrusion.
7. It could be abused by making employability conditional on having a user authentication number.
9. Downloading a file without virus checking turned on. Using obvious network login passwords or no password at all.

Index